Amazing World Records of Weather

Sunflower education

Exceptional Books for Teachers and Parents

A Great Way to Teach Weather!

Capture students' imagination with a jaw-dropping world record and then build on that interest to teach core ideas. For example, a lesson on the world's greatest rainfall (it's 73.62 inches—in one day!) leads to activities in which students construct rain gauges, record rainfall, and learn about the types of liquid precipitation. Students learn what makes weather, construct their own barometers, model the Coriolis Effect, apply the Beaufort scale, explain how snowflakes form, and conduct dozens of other high-interest activities.

Grades 5-9. Recommended by the National Science Teachers Association.
60 Reproducible Activity Sheets • Teaching Tips • Complete Answer Key

Part of the Amazing World Records Series of books!

WorldRecordsBooks.com

Please feel free to photocopy the activity sheets in this book within reason. Sunflower Education grants teachers permission to photocopy the activity sheets from this book for educational use. This permission is granted to individual teachers and not entire schools or school systems. For information or questions regarding permissions, please send an email to permissions@SunflowerEducation.net.

Visit **SunflowerEducation.Net** for more great books!

Editorial   Sunflower Education

Design   Cynthia Hannon Design

Photography

Cover and title page images: © iStockphoto LP.
Interior images: © Wikimedia Commons

ISBN-13: 978-1-937166-02-1
ISBN-10: 1-937166-02-3
Copyright © 2013
Sunflower Education
All rights reserved.
Printed in the U.S.A.

Contents

To the Teacher 2

WORLD RECORDS OF HOT AND COLD

1. The World's Sunniest Place 7
   A. The World's Sunniest Place 8
   B. What Makes Weather? 9
   C. The Uneven Heating of the Earth 11

2. The World's Hottest Place 12
   A. The World's Hottest Place 13
   B. The Dangers of Heat 14
   C. Folklore for Hot Days 16

3. The World's Highest Temperature 17
   A. The World's Highest Temperature 18
   B. Understanding Temperature Scales 19
   C. Converting Temperatures 20

4. The World's Coldest Place 22
   A. The World's Coldest Place 23
   B. The Dangers of Cold 24
   C. Folklore for Cold and Snow 26

5. The World's Lowest Temperature 27
   A. The World's Lowest Temperature 28
   B. Understanding Thermometers 29
   C. Tracking the Temperature 30

WORLD RECORDS OF AIR PRESSURE

6. The World's Highest Air Pressure 33
   A. The World's Highest Air Pressure 34
   B. What Is Air Pressure? 35
   C. Air Pressure and Weather 36

7. The World's Lowest Air Pressure 38
    A. The World's Lowest Air Pressure 39
    B. Understanding Barometers 40
    C. Making a Barometer 41

8. The World's Windiest Place 42
    A. The World's Windiest Place 43
    B. Why Does the Wind Blow? 44
    C. Understanding the Coriolis Effect 45

9. The World's Highest Wind Speed 46
    A. The World's Highest Wind Speed 47
    B. Wind and Its Effects 48
    C. Constructing a Weather Vane 50

WORLD RECORDS OF PRECIPITATION

10. The World's Wettest Place 53
    A. The World's Wettest Place 54
    B. Understanding Monsoons 55
    C. What Makes Rain? 56

11. The World's Greatest Rainfall 58
    A. The World's Greatest Rainfall 59
    B. Constructing a Rain Gauge 60
    C. Classifying Rain 61

12. The World's Greatest Snowfall 63
    A. The World's Greatest Snowfall 64
    B. Why Does It Snow? 65
    C. Snow 66

13. The World's Driest Place 67
    A. The World's Driest Place 68
    B. Rainmaking 69
    C. Understanding Drought 70

14. The World's Foggiest Place 71
    A. The World's Foggiest Place 72
    B. What Makes Fog? 73
    C. Ships in Fog 75

15. The World's Heaviest Hail 77
    A. The World's Heaviest Hail 78
    B. What Makes Hail? 79
    C. Huge Hailstones 80

WORLD RECORDS OF WEATHER DISASTERS

16. The World's Deadliest Hurricane 83
    A. The World's Deadliest Hurricane 84
    B. Understanding Hurricanes 85
    C. Tracking a Hurricane 87

17. The World's Worst Tornado 88
    A. The World's Worst Tornado 89
    B. Why Are There Tornadoes? 90
    C. Tornado Safety 91

18. The World's Worst Flood 92
    A. The World's Worst Flood 93
    B. The River of Sorrow 94
    C. Controlling Floods 96

19. The World's Worst Weather Disaster 97
    A. The World's Worst Weather Disaster 98
    B. Natural Disasters at the Head of the Bay of Bengal 99
    C. Preparing for a Natural Disaster 101

20. The World's Worst Weather 102
    A. The Country with the World's Worst Weather 103
    B. Understanding Weather in the United States 104
    C. The National Weather Service 105

To the Teacher

Amazing World Records of Weather is one in a series of books that explore the superlatives of a variety of subjects. Other titles in the series include *Amazing World Records of Geography, Amazing World Records of History, Amazing World Records of Language and Literature,* and *Amazing World Records of Science and Technology.*

Amazing World Records of Weather welcomes students to the superlatives of weather. In these pages they—and you—will discover the hottest, coldest, windiest, foggiest, wettest, driest, and most interesting and intense weather on Earth.

Topic Coverage

The dynamics of weather are extremely complex. Dozens of factors combine to produce the simplest weather phenomenon. These factors are simplified in this book to provide students with a general, level-appropriate introduction to weather. The basics are well covered, providing answers to the most common questions that students ask. This collection is an attempt to gather information about weather that is particularly interesting to students, useful as a gateway to discussion of broader meteorological topics, germane to general student learning, and high in pedagogical value.

How This Book Is Organized

Amazing World Records of Weather is a supplementary book, conceived and designed for you to access at your discretion. It also works well as a stand-alone text to introduce students to many of the major basic concepts of weather. The book is organized to make both options viable. If you choose the latter course, guide students as they use the book progressively.

Units

The world records are organized into four general units: World Records of Hot and Cold, World Records of Air Pressure, World Records of Precipitation, and World Record Weather Disasters. Each unit includes from four to six chapters.

Chapters

Each chapter addresses a particular world record. Each chapter opens with a Teacher Guide Page, which provides objectives, teaching tips, answers, and extension and enrichment activities. Each chapter includes three, 1-2-page reproducible student activity sheets.

Student Activity Sheets

The first student activity sheet always introduces the world weather record. The following two activity sheets vary in format, content, and student activity. In general, they expand on the record setter to introduce students to general aspects of that weather phenomenon.

How to Use This Book

The student activity sheets are at the heart of *Amazing World Records of Weather*. Virtually all of them can be used as the basis for individual, partner, group, or whole-class activities. They are self-explanatory. Your judgment is the central guide for how to best use them in your classroom. For further instruction, consult the information presented on the Teacher Guide Page.

The first student activity sheet in each chapter includes two features, "From the *Meteorologist's Handbook*" and "From the *World Weather Records Book*," which introduce the relevant terms and the world weather record. As students progress through the book, you may elect to have students clip these features out of the pages, paste them into student-made booklets, and thus create their own copies of a *Meteorologist's Handbook* and a *World Weather Records Book*.

The student activity sheets that are subtitled "People and the Weather" focus students on the essential interaction that makes weather so important. These address a variety of topics, from weather folklore to safety during extreme weather conditions.

Another group of activity sheets, all subtitled, "Studying Your Local Weather," guide students as they create basic weather instruments and record basic weather information. Students construct a barometer, a weather vane, and a rain gauge, and keep air pressure, temperature, wind, and precipitation logs. You might wish to group these activity sheets as you guide students in creating and utilizing a school weather station.

However you integrate *Amazing World Records of Weather* into your teaching; always keep in mind that the student activity sheets are perfect springboards to broader and deeper learning about weather. Keep in mind, too, that students learn best when they're having fun. Children, like everyone else, are naturally interested in the weather. Expand on their curiosity to give them a broader understanding. Emphasize the power, beauty, and sublimeness of the weather, and foster in your students both respect and wonder at something that will affect them every day of their lives.

World Records of Hot and Cold

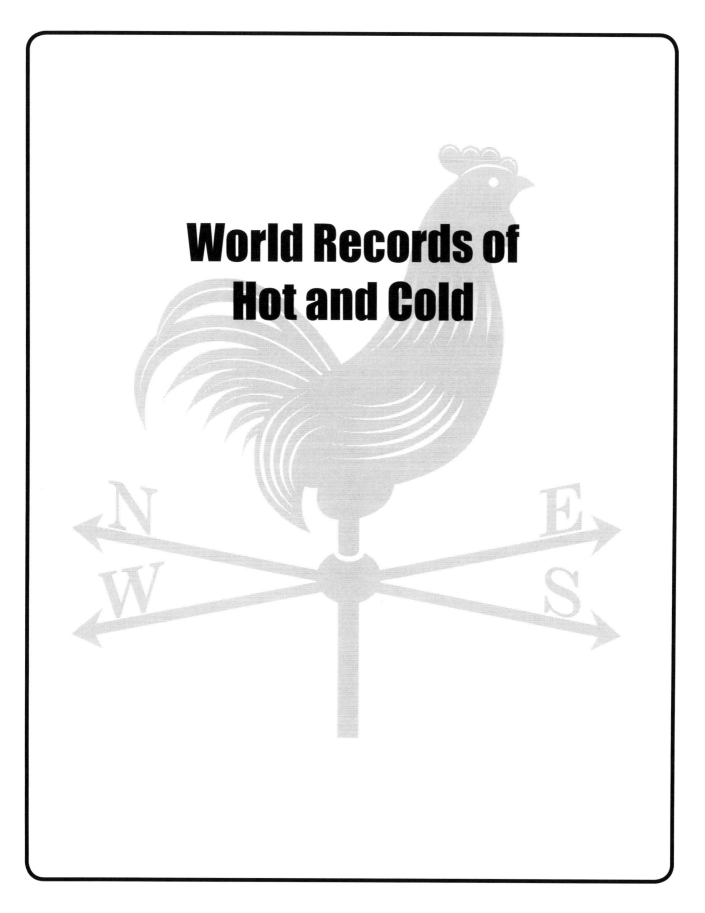

1. The World's Sunniest Place
Yuma, Arizona

Objectives
- Identify and discuss the world's sunniest place
- Explain the role of the Sun in creating Earth's weather
- Demonstrate how the Sun heats the Earth unevenly

Time and Special Materials
- About one class period
- Globe, flashlight, pen

Teaching Tips
You may wish to introduce the topic by completing the first activity sheet as a class. The two additional activity sheets can be completed by students working independently, with partners, or in small groups.

Activity Sheet 1A
- Check student comprehension of the terms in the *Meteorologist's Handbook*; you may have students clip the entries from the *Meteorologist's Handbook* and the *World Weather Records Book* and add them to their compilations.
- Focus students' attention on the map.

Activity Sheet 1B
- Emphasize the critical role of the Sun as the "fuel" that drives the "weather engine" of the Earth.
- Since this basic concept underlies all weather, take special care to ensure students understand it.

Activity Sheet 1C
- You may wish to perform this activity as a class-wide demonstration.
- Explain how the shadow cast by the pen demonstrates that the rays from the Sun hit the Earth at different angles at different latitudes.

Answers
- **Activity Sheet 1A** 1. About 91%; 2. There's no sunshine at night; 3. Answers will vary and may include the following responses: most people have tans, it doesn't rain very much, it is hot in Yuma. Reward thoughtful responses.
- **Activity Sheet 1B** 1. the Sun; 2. about one two-billionth; 3. the atmosphere; the different heating patterns around the world caused by the Sun; 4. The "weather engine" is the uneven heating of the Earth by the Sun and the resulting different air masses that, combined with moisture and the Earth's rotation, move, collide, and make weather.
- **Activity Sheet 1C** Have students work together on this activity. Reward honest effort.

Extension and Enrichment
- You may wish to explain how the tilt and the revolution of the Earth cause seasons.
- Have students hold the flashlight very close to the globe to see how the same amount of light covers a smaller area at the Equator and a larger area at higher latitudes.

Visit WorldRecordsBooks.com for more images and activities!

The World's Sunniest Place
Activity Sheet 1A

Name _____ Class _____ Date _____

FOCUS *Read the following entries from the* Meteorologist's Handbook *and from the* World Weather Records Book. *As you read, compare what you are reading with any personal experiences you have had with this type of weather. Then complete the activities that follow.*

From the *Meteorologist's Handbook*
- **weather:** the atmospheric conditions of a given location over a specific period of time
- **climate:** the prevailing or average weather conditions of a location or region over a longer period of time
- **sunshine:** the sunlight heating and lighting a particular location of the Earth's surface

From the *World Weather Records Book*
The location that receives the most sunshine is the city of Yuma, Arizona. On average, Yuma receives 4,055 hours of sunshine each year, out of a possible 4,456 hours.

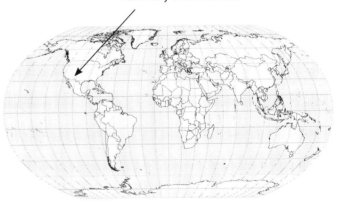

Yuma, Arizona

1. Study the statistics given in the *World Weather Records Book*. What percentage of possible hours of sunshine does Yuma receive each year?

   _____

2. There are about 8,760 hours in the year. But according to the *World Weather Records Book*, there are only 4,456 possible hours of sunshine. How do you account for this difference?

   _____
   _____

3. How do you think living in the world's sunniest place affects life in Yuma?

   _____
   _____

What Makes Weather?
Activity Sheet 1B

Name _____ Class _____ Date _____

FOCUS *What makes weather? The answer may surprise you, because what makes weather on Earth isn't on Earth at all—it's 93 million miles away! To see how this is so, read the article. Then answer the questions.*

The Sun and the Weather

You know that a sunny day is nice, and that the sunshine will warm you up. But did you know that the Sun also causes snow and rain? In fact, *the Sun causes all weather on the planet.*

How can this be? The Sun is a huge energy source. It's gigantic—more than a million times larger than the Earth—and extremely hot. At the Sun's surface, the temperature is about 10,000°F. Energy in the form of light and heat is constantly given off by the Sun into space.

Only about one two-billionth of the Sun's energy reaches the Earth. And much of the energy that does reach our planet is reflected back into space by clouds. But, because the Sun generates so much energy, the tiny fraction that reaches us is powerful enough to have enormous effects. The energy that reaches us is absorbed by the **atmosphere**—the layer of air surrounding the Earth—and by the ground and water. The atmosphere traps this energy next to the Earth. All of this energy provides "fuel" for the "weather engine" that causes the weather on Earth.

The Sun does not heat the Earth evenly. This is because the Earth is a sphere and because land and water absorb and release heat at different rates. At different places on Earth and at different times of the year, different amounts of the Sun's energy fall, heating the Earth and the atmosphere unevenly.

This uneven heating of the Earth causes changes in the atmosphere. Huge, three-dimensional areas of the atmosphere called **air masses** form. These areas have different temperatures and air pressures. Air masses of opposite temperatures (hot or cold) and opposite air pressures (high or low) move toward each other. These movements of the air combine with other factors (like the rotation of the Earth and the amount of water in the air) to make—you guessed it—our weather.

The point at which air masses collide is called a **front**. The fronts make weather. The Sun's energy strikes the Earth continuously, so the "weather engine" is always running. And it all starts with the energy of the Sun, 93 million miles from Earth.

1. What is the root cause of weather on Earth?

   _____
   _____
   _____

2. How much of the Sun's energy reaches the Earth?

   _____
   _____
   _____

3. What traps the Sun's energy near the surface of the Earth?

   _____
   _____
   _____

4. What is the "weather engine"?

   _____
   _____
   _____
   _____
   _____
   _____
   _____

The Uneven Heating of the Earth
Activity Sheet 1C

Name _____ Class _____ Date _____

FOCUS *The Sun creates all the weather on Earth, largely because the Sun heats the Earth unevenly. But why is the Earth heated unevenly? To find out, complete the following steps.*

____ **Step 1 Get a Globe and a Flashlight.** The globe will serve as a model of the Earth. The flashlight will represent the rays of light from the Sun.

____ **Step 2 Set Up Your Materials.** Make sure that the globe is tilting at about a 23 1/2° angle. (This is the same angle that the Earth is tilted in relation to the Sun.) Turn the flashlight on and position it so that it shines on the globe. The light should be aimed at the equator.

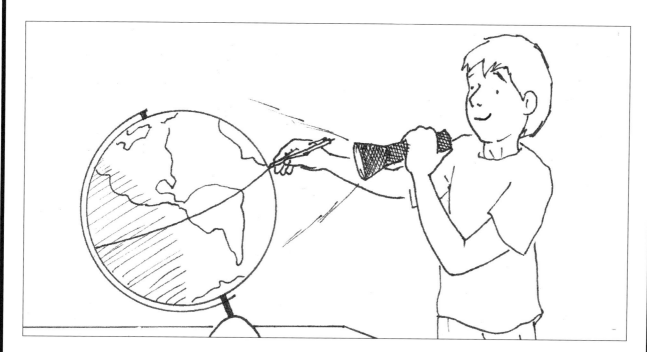

____ **Step 3 Measure the Sunlight.** Hold a pen at a right angle to the Earth on the equator with the other end of the pen directly facing the Sun. What kind of shadow is cast? Now hold the pen at a right angle to the "Earth" about halfway between the Equator and the North Pole. What kind of shadow is cast? What can you conclude about the angle of the Sun's rays at different locations on the Earth?

____ **Step 4 Think About What You've Seen.** How does your experiment show that the Sun heats the Earth unevenly?

2. The World's Hottest Place
Marble Bar, Australia

Objectives
- Identify and discuss the world's hottest place
- Discuss the dangers of heat
- Analyze weather folklore related to heat

Time Required
- About one class period

Teaching Tips
You may wish to introduce the topic by completing the first activity sheet as a class. The two additional activity sheets can be completed by students working independently, with partners, or in small groups.

Activity Sheet 2A
- Check student comprehension of the term in the *Meteorologist's Handbook*; you may have students clip the entries from the *Meteorologist's Handbook* and the *World Weather Records Book* and add them to their compilations.
- Focus students' attention on the map.

Activity Sheet 2B
- Emphasize the very real dangers of heat.
- Foster student awareness of the symptoms or warnings signs of each malady.

Activity Sheet 2C
- Explain that scientists have recently validated much weather folklore.
- Make sure students understand the origins of weather folklore and that it developed over decades and centuries.

Answers
- *Activity Sheet 2A* 1. Marble Bar in Australia; 2. Possible responses may include fewer people live there, animals are mainly nocturnal, and plants store water. Reward thoughtful responses.
- *Activity Sheet 2B* 1. heatstroke; 2. lie down in a cooler location; drink cold, salted water; 3. avoid exercise when the weather is too hot; drink plenty of water; and take it easy during hot weather.
- *Activity Sheet 2C* 1. Clouds cover the sky at night, and daytime clouds are hazy. The sunset is a bright red color, and the night air doesn't cool off quickly; 2. bugs drone loudly; cats stretch out full length; crickets start chirping more quickly; 3. Answers will vary. Reward thoughtful responses; 4. Answers will vary. Reward thoughtful responses.

Extension and Enrichment
- Consider having students interview appropriate adults about local weather folklore.
- If possible, have students test the idea that the chirping rate of crickets accurately reflects the temperature.

Visit WorldRecordsBooks.com for more images and activities!

The World's Hottest Place
Activity Sheet 2A

Name _____ Class _____ Date _____

FOCUS *Read the entries from the Meteorologist's Handbook and from the World Weather Records Book. As you read, compare what you are reading with any personal experiences you have had with this type of weather. Then answer the questions that follow.*

From the *Meteorologist's Handbook*
- **temperature:** the amount of heat in the Earth's atmosphere. This heat comes from the Sun.

From the *World Weather Records Book*
The **world's hottest place** is Marble Bar in Australia. The highest temperature ever recorded there was 120.5° F. However, this is not the highest temperature ever recorded. What makes Marble Bar the world's hottest place is its continuously high temperatures. Between October, 1923 and April, 1924, Marble Bar experienced 160 consecutive days—more than five months—when the temperature soared above 100° F.

1. What is the hottest place on Earth?

   _____
   _____

2. How do you think the heat affects human, plant, and animal life in this area?

   _____
   _____

Marble Bar, Australia

_____
_____
_____
_____
_____

SunflowerEducation.net   Amazing World Records • Weather

The Dangers of Heat
Activity Sheet 2B

Name _____ Class _____ Date _____

FOCUS *Without heat, there would be no human life. But too much heat can be harmful—even deadly. The residents of Marble Bar, Australia—the hottest place on Earth—are keenly aware of these dangers. So should you be. To learn about the dangers of heat and what to do about them, study the chart and answer the questions.*

1. What is the most serious heat-related danger?

   _____

   _____

2. What should be done to treat heat exhaustion?

   _____

   _____

   _____

   _____

3. What three steps should you always take to prevent heat-related illnesses?

   _____

   _____

   _____

   _____

   _____

   _____

   _____

Dangers of the Heat

Condition	Symptoms and Causes	Prevention and Treatment
Heat Cramps	**Symptoms** • cramps, especially in the legs **Causes** • exercising in hot weather leading to an imbalance of body salts	**Prevention** • avoid exercising when the weather is too hot • drink plenty of water • take it easy during hot weather **Treatment** • go to a cooler location • drink water • rest
Fainting	**Symptoms** • loss of consciousness **Causes** • exercising in hot weather leading to a drop in blood pressure	**Prevention** • avoid exercising when the weather is too hot • drink plenty of water • take it easy during hot weather **Treatment** • go to a cooler location • drink water • rest
Heat Exhaustion	**Symptoms** • cramps, dizziness, weakness, headache, nausea; face looks pale; rapid pulse, shallow breathing; skin feels cold and moist **Causes** • inadequate body fluids and salts	**Prevention** • avoid exercising when the weather is too hot • drink plenty of water • take it easy during hot weather **Treatment** • lie down in a cooler location • drink cold, salted water
Heatstroke	**Symptoms** • headache, dizziness, and nausea; signs of confusion, lack of coordination, delirium, then unconsciousness; skin feels hot and dry; severe increase in body temperature **Causes** • extreme heat couples with high humidity to impair the body's cooling system *Heatstroke can cause death.*	**Prevention** • avoid exercising when the weather is too hot • drink plenty of water • take it easy during hot weather **Treatment** • move victim to a cool location • cool the victim by any means possible (cover with a wet sheet, fan) • give fluids when the victim can drink them • *Seek medical help immediately when heatstroke is suspected.*

Folklore for Hot Days
Activity Sheet 2C

Name _____ Class _____ Date _____

FOCUS — *Thousands of years before modern weather forecasts, people knew a lot about the weather simply by observing it day to day. The knowledge they gathered has been passed down from generation to generation. Such common knowledge of the weather by everyday "folks" is called "weather folklore" or simply weather lore. Common weather lore that supposedly signals a hot day is listed in the box. Read and think about each one. Then answer the questions.*

Signs That The Day Will Be Hot

The day will be hot if...
- clouds cover the sky at night and daytime clouds are hazy.
- the sunset is a bright red color and the night air doesn't cool off quickly.
- bugs are droning especially loudly.
- cats stretch out full length.
- crickets start chirping more quickly. (Crickets chirp faster at higher temperatures. By counting the number of chirps in 15 seconds, and adding 37, you can determine the temperature.)

1. Which weather signs seem to be based on observations of the atmosphere?

2. Which ones are based on the behavior of animals?

3. Do any of these seem to make sense to you? Explain.

4. Do any of them seem farfetched? Explain.

3. The World's Highest Temperature
136° Fahrenheit

Objectives
- Identify and discuss the world's highest temperature
- Compare and contrast the Fahrenheit and Celsius temperature scales
- Convert Fahrenheit temperatures to Celsius temperatures and vice versa

Time and Special Materials
- About one class period
- Calculators

Teaching Tips
You may wish to introduce the topic by completing the first activity sheet as a class. The two additional activity sheets can be completed by students working independently, with partners, or in small groups.

Activity Sheet 3A
- Check student comprehension of the term in the *Meteorologist's Handbook*; you may have students clip the entries from the *Meteorologist's Handbook* and the *World Weather Records Book* and add them to their compilations.
- Focus students' attention on the map.

Activity Sheet 3B
- Explain how the Celsius scale is a metric scale that is widely used throughout the world.
- Point out how both scales are based on the behavior of water at different temperatures.

Activity Sheet 3C
- You may wish to allow students to use calculators to complete this activity.
- Explain why knowing how to make these conversions is important.

Answers
- *Activity Sheet 3A* 1. 136°; 2. Al'Aziziyah, Libya; 3. Answers will vary. Reward thoughtful responses.
- *Activity Sheet 3B* 1. Fahrenheit 32°, 212°; Celsius 0°, 100°; 2. Fahrenheit; 3. Centigrade is from the French "centi" for "hundred" and the Latin gradus or degree.
- *Activity Sheet 3C* Check the charts to ensure the correct conversions.

Extension and Enrichment
- Display a thermometer that has both Fahrenheit and Celsius scales.
- Have students find out about the Kelvin scale and compare and contrast it to the Fahrenheit and Celsius scales.

Visit WorldRecordsBooks.com for more images and activities!

The World's Highest Temperature
Activity Sheet 3A

Name _____ Class _____ Date _____

FOCUS *Read the entries from the* Meteorologist's Handbook *and from the* World Weather Records Book. *As you read, compare what you are reading with any personal experiences you have had with this type of weather. Then answer the questions that follow.*

From the *Meteorologist's Handbook*

- **temperature scale:** system for measuring the amount of heat in the atmosphere, using a thermometer. The two most common temperature scales are the Fahrenheit (F) scale and the Celsius (C) scale.

From the *World Weather Records Book*

The **world's highest temperature** was recorded on September 13, 1922, at Al'Aziziyah, Libya. The thermometer registered a stunning 136° F—in the shade.

1. What was the highest temperature ever recorded?

   _____

2. Where was this temperature recorded?

   _____
   _____

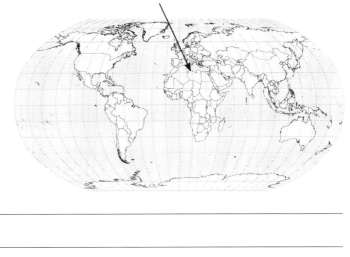

Al'Aziziyah, Libya

3. What is the highest temperature you've ever experienced? What did it feel like?

   _____
   _____
   _____
   _____
   _____

Understanding Temperature Scales

Activity Sheet 3B

Name _____ Class _____ Date _____

FOCUS *Will it surprise you to learn that the highest temperature ever recorded was a sweltering 58 degrees? It won't if you understand that temperature is measured in different ways. These different ways are discussed in the article. Read the article and answer the questions.*

Two Temperature Scales

Temperatures can be measured with different scales. The two most common are the Fahrenheit scale and the Celsius scale. (Another scale, the Kelvin scale, is used in scientific measurements.)

If you live in the United States, you're probably most familiar with the Fahrenheit scale, because that's the one used in this country. When the temperature is a pleasant 72 degrees, for example, that's 72 degrees Fahrenheit. Sometimes, to make it clear, we use the abbreviations F or C to indicate which scale is being used.

Both the Fahrenheit and the Celsius scales are based on how water behaves at different temperatures. On the Fahrenheit scale, water freezes at 32 degrees and boils at 212 degrees. On the Celsius scale, water freezes at 0 degrees and boils at 100 degrees.

Where do we get the names for these scales? Fahrenheit was the name of a scientist, Gabriel Fahrenheit, who made a mercury thermometer way back in 1714. That is the type of thermometer we use today. Celsius is named for Anders Celsius, the scientist who devised a scale similar to the modern Celsius in 1742.

1. Complete the chart below.

Temperature Scale	Temperature at Which Water Freezes	Temperature at Which Water Boils
Fahrenheit		
Celsius		

2. Which scale is used more commonly in the United States?

_____

3. The Celsius scale is also called the centigrade scale. Conduct research and explain why this is so.

_____

Converting Temperatures
Activity Sheet 3C

Name _____ Class _____ Date _____

FOCUS *Although the United States uses the Fahrenheit scale, much of the rest of the world uses the Celsius scale. If you ever travel abroad, or read about foreign countries, you'll need to convert one scale to the other. On this page, you'll learn how to do this, and you'll practice what you've learned. Study the information in the boxes. Then complete the activities.*

To Convert a Fahrenheit Temperature to a Celsius Temperature

1. Subtract 32 from the Fahrenheit temperature.
2. Divide the result by 1.8 to get the Celsius temperature.

Example

Fahrenheit Temperature: 72 degrees

1. 72 - 32 = 40
2. 40 / 1.8 = 22

72 degrees Fahrenheit equals 22 degrees Celsius.

To Convert a Celsius Temperature to a Fahrenheit Temperature

1. Multiply the Celsius temperature by 1.8.
2. Add 32 to the result to get the Fahrenheit temperature.

Example

Celsius Temperature: 30 degrees

1. 30 x 1.8 = 54
2. 54 + 32 = 86

30 degrees Celsius equals 86 degrees Fahrenheit.

Converting Temperatures

Complete the chart. Round off your answers to the nearest degree.

Temperature in Degrees Fahrenheit	Temperature in Degrees Celsius
0	
20	
	0
	5
60	
72	
	30
	40
136 (the highest temperature ever recorded)	
	100

This thermometer has both the Fahrenheit and Celsius scale.

4. The World's Coldest Place
Polyus Nedostupnosti, Antarctica

Objectives
- Identify and discuss the world's coldest place
- Describe and discuss the dangers of extreme cold
- Analyze weather folklore related to the cold

Time Required
- About one class period

Teaching Tips
You may wish to introduce the topic by completing the first activity sheet as a class. The two additional activity sheets can be completed by students working independently, with partners, or in small groups.

Activity Sheet 4A
- Check student comprehension of the terms in the *Meteorologist's Handbook*; you may have students clip the entries from the *Meteorologist's Handbook* and the *World Weather Records Book* and add them to their compilations.
- Focus students' attention on the map.

Activity Sheet 4B
- Emphasize the very real dangers of cold.
- Foster student awareness of the dangers of cold weather.

Activity Sheet 4C
- Explain how scientists have recently validated much weather folklore.
- Make sure students understand the origins of weather folklore and that it has been developed over decades and centuries.

Answers
- *Activity Sheet 4A* 1. Answers will vary and may include the following responses: yes, I would have expected the coldest place to be at the South Pole. Reward thoughtful responses; 2. The angles of the Sun's rays are low at the poles.
- *Activity Sheet 4B* 1. the freezing of skin due to extreme cold; 2. a serious drop in body temperature; 3. seek immediate medical attention; 4. Be aware of the dangers of cold weather. Take care to dress warmly and keep extremities covered. Avoid going out or staying out for too long in cold weather.
- *Activity Sheet 4C* 1. A clear moon means a frost soon. A heavy snowfall is always preceded by three cloudy days; 2. Cows only moo in the evening if cold air and snowy weather are on the way; 3. Answers will vary. Reward thoughtful responses; 4. Answers will vary. Reward thoughtful responses.

Extension and Enrichment
- You may wish to have students dramatize the symptoms and treatments of cold-related dangers.
- Invite students to share memories of cold spells in your community.

Visit WorldRecordsBooks.com for more images and activities!

The World's Coldest Place
Activity Sheet 4A

Name _____ Class _____ Date _____

FOCUS *Read the entries from the* Meteorologist's Handbook *and from the* World Weather Records Book. *As you read, compare what you are reading with any personal experiences you have had with this type of weather. Then answer the questions that follow.*

From the *Meteorologist's Handbook*
- **thermometer:** an instrument that measures temperature

From the *World Weather Records Book*
The **world's coldest place** is Polyus Nedostupnosti, Antarctica. This is not the site of the coldest temperature ever recorded. What makes Polyus Nedostupnosti the coldest place is that it's bitterly cold there all the time. The temperature there averages -72° F.

1. Does the location of the world's coldest place surprise you? Why or why not?

   _____
   _____
   _____

2. Explain how the angles of the Sun's rays at Polyus Nedostupnosti help make it the coldest place on Earth.

   _____
   _____
   _____
   _____

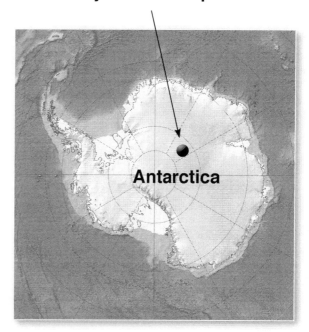

The Dangers of Cold
Activity Sheet 4B

Name _____ Class _____ Date _____

FOCUS *All of the people who visit Antarctica are keenly aware of the cold. They are also aware of the dangers of cold weather. Because these dangers can occur at surprisingly high temperatures, you need to be aware of these dangers, too. Study the chart and answer the questions.*

Dangers of the Cold

Condition	Definition and Description	Preventative Measures
Frostbite	• Frostbite is the freezing of skin due to extreme cold. • Ice crystals form in the skin and damage tissue. • Extremities such as hands, feet, ears and the nose are particularly vulnerable. • *Frostbite may require amputation.*	• Be aware of the dangers of cold weather. • Take care to dress warmly and keep extremities covered. • Avoid going out or staying out for too long if the weather is cold. • *Always seek immediate medical attention if frostbite is even suspected.*
Hypothermia	• Hypothermia is a serious drop in body temperature. • *It can be deadly.*	• Be aware of the dangers of cold weather. • Take care to dress warmly and keep extremities covered. • Avoid going out or staying out for long if the weather is cold. • *Always seek immediate medical attention if hypothermia is even suspected.*

1. What is frostbite?

   _____

   _____

2. What is hypothermia?

   _____

   _____

3. What should you always do if you suspect either of these conditions?

   _____

   _____

4. What three steps should you always take to protect yourself from these conditions?

   _____

   _____

   _____

Folklore for Cold and Snow
Activity Sheet 4C

Name _____ Class _____ Date _____

FOCUS — *Thousands of years before modern weather forecasts, people knew a lot about the weather simply by observing it from day to day. The knowledge they gathered has been passed down from generation to generation. Such common knowledge of the weather by everyday "folks" is called weather folklore or simply weather lore. Some weather lore that supposedly signals the onset of cold weather is listed in the box. Read and think about each one. Then answer the questions.*

Signs That a Spell of Cold or Snow is Coming
- A clear moon means a frost soon.
- A heavy snowfall is always preceded by three cloudy days.
- If smoke hangs low in the sky and doesn't dissipate, it is about to snow.
- Cows only moo in the evening if cold air and snow are on the way.
- The needles of pine trees point west before a snowfall.

1. Which weather signs seem to be based on observations of the atmosphere?

_____
_____
_____

2. Which one is based on the behavior of animals?

_____

3. Do any of these seem to make sense to you? Explain.

_____
_____

4. Do any of them seem farfetched? Explain.

_____
_____
_____

5. The World's Lowest Temperature
-128.6° Fahrenheit

Objectives
- Identify and discuss the world's lowest temperature
- Explain how thermometers function
- Record and analyze local temperatures over an extended period of time.

Time and Special Materials
- About one class period
- Thermometer(s)

Teaching Tips
You may wish to introduce the topic by completing the first activity sheet as a class. The two additional activity sheets can be completed by students working independently, with partners, or in small groups.

Activity Sheet 5A
- Check student comprehension of the term in the *Meteorologist's Handbook*; you may have students clip the entries from the *Meteorologist's Handbook* and the *World Weather Records Book* and add them to their compilations.
- Focus students' attention on the map.

Activity Sheet 5B
- Explain why there is a vacuum at the top of a liquid-glass thermometer.
- Have students identify places where each type of thermometer is used.

Activity Sheet 5C
- This activity requires substantial teacher guidance.
- Rotate the responsibility for keeping the temperature log among students.

Answers
- *Activity Sheet 5A* 1. -128.6° F; 2. Answers will vary. Reward thoughtful responses; 3. Answers will vary.
- *Activity Sheet 5B* 1. Changes in temperature affect the way certain substances behave; 2. liquid-glass, deformation-type, electrical; 3. liquid-glass thermometers.
- *Activity Sheet 5C* Assist students in setting up the thermometer in a location where it won't be disturbed. Take regular readings. Have students work out the average temperature.

Extension and Enrichment
- Have students compare their temperature readings with average temperature readings for your community.
- Have students integrate their temperature logs into a climate profile of your community.

Visit WorldRecordsBooks.com for more images and activities!

The World's Lowest Temperature
Activity Sheet 5A

Name _____ Class _____ Date _____

FOCUS *Read the entries from the* Meteorologist's Handbook *and from the* World Weather Records Book. *As you read, compare what you are reading about with any personal experiences you have with this type of weather. Then answer the questions that follow.*

From the *Meteorologist's Handbook*
• **log:** a written, dated record of weather measurements

From the *World Weather Records Book*
The **world's lowest temperature** was recorded at Vostok, Antarctica, on July 21, 1983. The thermometer registered an almost unbelievably low temperature of -128.6 degrees Fahrenheit.

1. What was the lowest temperature ever recorded?

   _____

2. Describe the time when you felt the coldest.

   _____
   _____
   _____
   _____
   _____
   _____
   _____

3. What was the approximate temperature then?

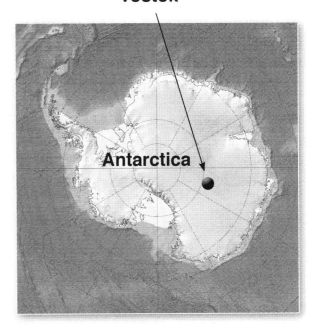

Understanding Thermometers
Activity Sheet 5B

Name _____ Class _____ Date _____

FOCUS *A thermometer is simply a device that measures temperature. You probably use thermometers all the time. You have them in your home and at school. But how do thermometers work? Read the explanation. Then answer the questions.*

How Thermometers Work

Thermometers work because changes in temperature affect the way certain substances behave. For example, when a liquid is heated, its volume expands. Since scientists have learned exactly how much liquids expand in relation to temperature, thermometers can be built.

The most common thermometers use liquid to measure temperatures. These are called *liquid-glass* or *liquid-in-glass* thermometers. You're the most familiar with these. In a liquid glass thermometer, the liquid is contained in a glass bulb at the bottom of a sealed glass pipe. As the temperature increases, the volume of the liquid increases, and the liquid travels up the pipe. A scale is printed along the pipe so we can read the exact temperature. Usually, the liquid used is mercury, because it expands and contracts evenly and remains liquid over a wide range of temperatures. Liquid-glass thermometers are commonly used for measuring air and body temperatures and in cooking.

Deformation-type thermometers deform, or change shape, in reaction to temperature changes. Strips of metal are often used in these thermometers. As the temperature increases or decreases, the metal expands or contracts, moving a pointer along a scale. Deformation-type thermometers are widely used in thermostats.

Electrical thermometers use electric currents to measure changes in temperature. The change in temperature causes a change in the flow of electricity, which is translated into a reading.

1. What principle underlies the way every type of thermometer works?

   _____

   _____

2. What are three main types of thermometers?

   _____

3. What is the most common type of thermometer?

   _____

Tracking the Temperature
Activity Sheet 5C

Name _____ Class _____ Date _____

FOCUS *The world's hottest place (Marble Ball, Australia: temperature commonly over 100 degrees Fahrenheit) and the world's coldest place (Polyus Nedostupnosti, Antarctica: average temperature -72 degrees Fahrenheit) didn't win their world weather records by being extremely hot or cold over the course of just one day or even over a few days. They won them by having extreme average temperatures over a long period of time. How does the average temperature of your school compare with these world records? Complete the following steps to find out. Check off each step as you complete it.*

____ **Step 1** Set Up a Thermometer.
Your teacher will guide you in finding a suitable location for your thermometer.

____ **Step 2** Keep a Temperature Log.
Note the temperature at the same time each day, and write it down in a temperature log. (Your teacher will help you compensate for days when you are not at school.) Your log should look something like this:

Date	Temperature (in degrees Fahrenheit)
April 7	72
April 8	74

____ **Step 3** Determine Average Temperature.
Keep your log for as long as you can. Determine the average weekly, monthly, and seasonal temperatures. How does the average temperature at your school compare to the average temperature of the hottest and coldest places in the world?

_____
_____
_____
_____
_____

World Records of Air Pressure

6. The World's Highest Air Pressure
32 inches

Objectives
- Identify and discuss the world's highest air pressure
- Define and explain air pressure
- Explain the role that air pressure plays in creating weather

Time and Required
- About one class period

Teaching Tips
You may wish to introduce the topic by completing the first activity sheet as a class. The two additional activity sheets can be completed by students working independently, with partners, or in small groups.

Activity Sheet 6A
- Check student comprehension of the terms in the *Meteorologist's Handbook*; you may have students clip the entries from the *Meteorologist's Handbook* and the *World Weather Records Book* and add them to their compilations.
- Focus students' attention on the map.

Activity Sheet 6B
- Emphasize that air is a real substance with volume and weight.
- Understanding that air pressure is vital, yet achieving this understanding can be difficult for students; be patient and work diligently to achieve student comprehension.

Activity Sheet 6C
- Have students memorize the two basic facts in the article.
- Stress the fundamental role of the uneven heating of the Earth by the Sun.

Answers
- *Activity Sheet 6A* 1. a barometer; 2. inches on the barometric scale; 3. 32 inches; 4. 107% of the average barometric pressure.
- *Activity Sheet 6B* 1. atmosphere; 2. The weight of air molecules is exerted in all directions creating air pressure; 3. Answers will vary. Reward thoughtful responses.
- *Activity Sheet 6C* 1. Air moves from high-pressure locations to low-pressure areas; 2. The Sun heats parts of the atmosphere, and warm air has lower air pressure than cool air; 3. Weather is generated when the air pressure in the Earth's atmosphere evens out.

Extension and Enrichment
- Have students report on how air pressure is an important part of local weather forecasts.
- Demonstrate the effect of air pressure by covering a glass that is about one-third filled with water with a paper plate and inverting it. The plate is held in place by air pressure.

Visit WorldRecordsBooks.com for more images and activities!

The World's Highest Air Pressure
Activity Sheet 6A

Name _____ Class _____ Date _____

FOCUS *Read the entries from the* Meteorologist's Handbook *and from the* World Weather Records Book. *As you read, compare what you are reading with any personal experiences you have had with this type of weather. Then complete the activities that follow.*

From the *Meteorologist's Handbook*

- **air pressure:** the force of the atmosphere pressing on the Earth. Air pressure is measured by an instrument called a *barometer* and expressed in *inches* on the barometer's scale. Because a barometer is used to measure air pressure, air pressure is also sometimes called *barometric pressure*.

From the *World Weather Records Book*

The **highest air pressure** ever recorded was 32 inches at Agata, Russia, on December 31, 1968.

1. What instrument is used to measure air pressure?

   _____

2. What unit is used to measure barometric pressure?

   _____

3. What was the highest barometric pressure recorded?

   _____

4. The sea-level average barometric pressure around the world is 29.92 inches. How does the highest-ever barometric pressure compare to this average? Express your answer as a percentage.

   _____

   _____

   _____

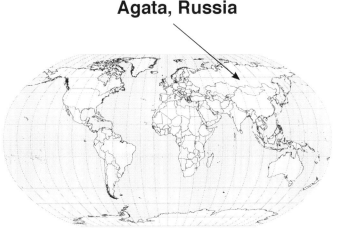

Agata, Russia

34 Amazing World Records • Weather

What Is Air Pressure?

Activity Sheet 6B

Name _____ Class _____ Date _____

FOCUS *Read the article about air pressure. Then answer the questions that follow.*

Air Pressure

You're under a lot of pressure right now, whether you know it or not.

The pressure is air pressure. You know that air is the mixture of gasses that surround the Earth. Oftentimes this is called the *atmosphere*. Air pressure is the force that the air exerts in all directions.

To understand air pressure, think about a book. If you lift up a book or any other solid object, you can easily feel its weight. You can also easily feel the weight of a liquid, such as water. Air, which is a gas, has weight too. But since it's much lighter than solids or liquids, you usually don't notice it.

How much does air weigh? A cubic foot of air weighs about one ounce.

The air pressure pushing down on you at this very moment from all of the air above you is about one ton. Why aren't you squished by it? Because air exerts pressure in all directions—down, up, and sideways. There is air pressure on you from all sides, supporting you and compensating for the pressure from above.

Scientists used to think that air pressure was caused by the great weight of the atmosphere pushing down. But we now know that air pressure is caused by the motion of air molecules. Air molecules move in all directions at incredible speeds, colliding with each other and anything else they touch. The impact of these billions of molecules causes the air to generate pressure. That's why air pressure is exerted in all directions. That's also why the gravity of the Earth doesn't pull all of the air in the atmosphere down to the Earth's surface.

1. What is another term for air?

   _____

2. How does the movement of molecules create air pressure?

   _____
   _____

3. Write a one-sentence definition of air pressure.

   _____

Air Pressure and Weather
Activity Sheet 6C

Name _____ Class _____ Date _____

FOCUS *Read the article. Then answer the questions.*

Differences in Air Pressure Help Cause the Weather

Differences in air pressure in different places in the atmosphere are a major reason why we have weather. Why? Because of two important facts about air pressure.

Fact: Air pressure "wants to be even."

To see what this means, imagine you have two airtight boxes. They're exactly the same size. But one has more air in it. This is because more air has been squeezed into the same space under higher pressure. If you suddenly connected the two boxes with a tube, air would rush from the high-pressure box to the low-pressure box, until the air pressure evens out in both boxes. (This is the same principle that gives you a flat tire on your bike. The air in your tire is at a higher pressure than the surrounding atmosphere. When something punctures the tire, the air from the high-pressure tire rushes out to the lower-pressure atmosphere, equalizing air pressure, but giving you a flat.)

Fact: Air pressure varies throughout the atmosphere.

At any given moment, there are thousands of areas of higher pressure and areas of lower pressure in the atmosphere. But if air pressure "wants to be even," then why doesn't the air pressure even out throughout the atmosphere and stay even?

This would happen if it weren't for the Sun. The Sun heats the Earth unevenly. In places where the atmosphere receives more of the Sun's energy, the air warms up. Warmer air has lower pressure than cooler air. The atmosphere is constantly trying to "even out" these different pressures. But just as soon as it's even, energy from the Sun creates more high and low-pressure areas, and the process continues. The result is that air is constantly moving from higher-pressure areas to lower-pressure areas.

As the air moves, it makes wind. It picks up and deposits moisture. It moves up and down. It creates and destroys clouds. In other words, it makes weather.

Of course, many other forces come into play to create weather. But, differences in air pressure, caused by differences in air temperature as a result of uneven heating from the Sun, are fundamental to weather on Earth.

1. What is meant by "air pressure wants to be even"?

2. How does the Sun create uneven air pressure?

3. How does this uneven air pressure help create weather?

7. The World's Lowest Air Pressure
25.69 inches

Objectives
- Identify and discuss the world's lowest air pressure
- Explain the operation of barometers
- Construct a simple barometer

Time and Special Materials
- About one class period
- Tall clear plastic glass, a long-necked, clear soda bottle, water
- Food coloring, marking pen, tape

Teaching Tips
You may wish to introduce the topic by completing the first activity sheet as a class. The two additional activity sheets can be completed by students working independently, with partners, or in small groups.

Activity Sheet 7A
- Check student comprehension of the terms in the *Meteorologist's Handbook*; you may have students clip the entries from the *Meteorologist's Handbook* and the *World Weather Records Book* and add them to their compilations.
- Focus students' attention on the map.

Activity Sheet 7B
- Remind students of the concept and meaning of air pressure.
- Explain why barometers have a vacuum at the top of the glass tube.

Activity Sheet 7C
- You may wish to have students cooperate to make a single classroom barometer.
- Guide students in comparing the performance and readings of their barometers with professional instruments and readings.

Answers
- *Activity Sheet 7A* 1. during a typhoon, in the Philippine Sea, in the southwestern Pacific Ocean; 2. a. a typhoon; b. extreme air pressure coincides with extreme weather; c. in a very dangerous location around a typhoon.
- *Activity Sheet 7B* 1. Evangelista Torriceli; 2. Because the first barometers measured the displacement of mercury in a glass "straw"; 3. Encourage students to visit the library to learn about aneroid barometers. Reward thoughtful responses.
- *Activity Sheet 7C* Assist students with constructing barometers. Reward honest effort.

Extension and Enrichment
- Have students report on how air pressure is an important part of local weather forecasts.
- Invite a local meteorologist to make a presentation on air pressure and answer students' questions.

Visit WorldRecordsBooks.com for more images and activities!

The World's Lowest Air Pressure
Activity Sheet 7A

Name _____ Class _____ Date _____

FOCUS *Read the entries from the* Meteorologist's Handbook *and from the* World Weather Records Book. *As you read, compare what you are reading with any personal experiences you have had with this type of weather. Then answer the questions that follow.*

From the *Meteorologist's Handbook*
- **atmosphere:** the envelope of gases surrounding the Earth; the air
  Air pressure is sometimes called *atmospheric pressure*.

From the *World Weather Records Book*
The **lowest air pressure** ever recorded was 25.69 inches during a typhoon on October 12, 1979, in the Philippine Sea, in the southwestern Pacific Ocean.

1. Describe the location at which the lowest air pressure was recorded.

   _____

2. Read the entry from the World Weather Records Book.

   a. What was happening when the pressure dropped so low?

   _____

   b. What does this tell you about the relationship between low air pressure and the weather?

   _____
   _____

   c. What can you conclude about where the barometer that measured this air pressure was located?

   _____
   _____
   _____

SunflowerEducation.net — Amazing World Records • Weather

Understanding Barometers
Activity Sheet 7B

Name _____ Class _____ Date _____

FOCUS *Air pressure is measured by a barometer and expressed in inches. But what exactly is a barometer? And how can air pressure be measured by a ruler? Read the article to find out. Then answer the questions that follow.*

Barometers

A barometer is a scientific instrument that measures air pressure. The barometer was invented in 1643 by a man named Evanelista Torricelli. Torricelli was one of Galileo's assistants.

It's easy to understand how barometers work. Just think about drinking a glass of water through a straw. When you suck on the straw, you suck the air out of it. The air pressure pushing down on the water in the glass pushes the water up into the space in the straw where the air was.

Barometers work in the same way. In a barometer, the "straw" is a glass tube, closed on top. The open bottom of the tube is placed in a container of mercury, just like a straw is placed in a glass of water. The air pressure pressing down on the mercury in the container forces some of the mercury up into the tube. By placing a ruler next to the tube, the distance the mercury flows up the tube can be measured in inches.

If the air pressure is lower, less force pushes down on the mercury in the container, and the mercury travels fewer inches up the tube. If the air pressure is higher, greater force pushes down on the mercury in the container, and the mercury travels more inches up the tube. Because the first barometers were constructed in this way, we measure air pressure in inches.

1. Who invented the barometer?
   _____

2. Why is air pressure measured in inches?
   _____

3. Conduct research to find out what an *aneroid barometer* is. Why are aneroid barometers now more common than mercury barometers?
   _____
   _____
   _____

Making a Barometer
Activity Sheet 7C

Name _____ Class _____ Date _____

FOCUS *To make a simple barometer, complete the following steps. Check off each one as you complete it.*

____ **Step 1 Gather Your Materials.** You'll need a tall clear plastic glass, a long-necked soda bottle with the label removed, food coloring, and a marking pen.

____ **Step 2 Construct Your Barometer.** Fill the glass of water about 1/3 full. Add a drop or two of food coloring to the water to help you see it. Next, turn the bottle upside down in the glass. You'll want the mouth of the bottle to hang just above the bottom of the glass. (You might need to shorten the glass so the bottle fits properly.)

____ **Step 3 Measure the Air Pressure.**
The force of the atmosphere pushing on the water in the glass will force some of the water up into the bottle. Use the tape to mark its level. Write the date on the tape. When the air pressure drops, the level of water in the bottle will drop. When the air pressure increases the water level will rise.

Since your barometer doesn't use mercury, the inches that the water travels up and down the bottle won't be the same as the inches in a standard barometer. But the way your barometer works—higher pressure increasing the level of the liquid, lower pressure decreasing it—is exactly the same as the way professional ones do.

____ **Step 4 Keep an Air Pressure Log.** Note the readings from your barometer in a daily air pressure log. Be sure to note the date of each reading.

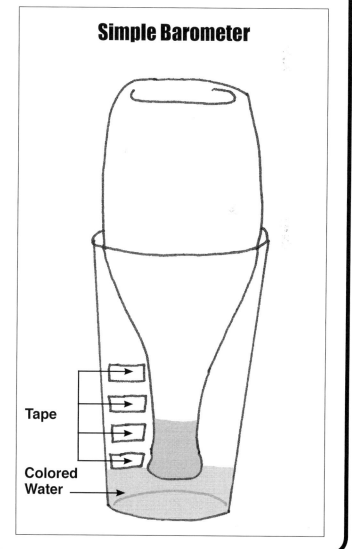

Simple Barometer

Tape

Colored Water

SunflowerEducation.net   Amazing World Records • Weather   **41**

8. The World's Windiest Place
Port Martin, Antarctica

Objectives
- Identify and discuss the world's windiest place
- Explain why the wind blows
- Investigate and demonstrate the Coriolis Effect

Time and Special Materials
- About one class period
- Lazy Susan, construction paper, tape, marking pen

Teaching Tips
You may wish to introduce the topic by completing the first activity sheet as a class. The two additional activity sheets can be completed by students working independently, with partners, or in small groups.

Activity Sheet 8A
- Check student comprehension of the terms in the *Meteorologist's Handbook*; you may have students clip the entries from the *Meteorologist's Handbook* and the *World Weather Records Book* and add them to their compilations.
- Focus students' attention on the map.

Activity Sheet 8B
- Students should memorize the rhyme, "Winds blow from high to low."
- Since this basic concept is fundamental to understanding weather, take special care to ensure student comprehension.

Activity Sheet 8C
- You may wish to perform this activity as a class-wide demonstration.
- Remind students that the Earth continuously rotates on its axis.

Answers
- *Activity Sheet 8A* 1. Port Martin, Antarctica; south of Australia; 2. Answers will vary and may include the following responses: the wind blows more of the time at sea; mountains block the wind; there is normally wind blowing off the sea. Reward thoughtful responses.
- *Activity Sheet 8B* 1. Different parts of the atmosphere have different air pressures. In order to even out the air pressure, the air moves from areas with high pressure to areas with low pressure, thus creating wind; 2. The pressure gradient force is the combination of the difference in air pressure and the distance apart; 3. The greater the PGF, the greater the wind speed.
- *Activity Sheet 8C* Assist students with the activity. Reward honest effort.

Extension and Enrichment
- You may wish to teach a brief lesson on worldwide wind patterns and how they are determined by air pressure and the Coriolis Effect.
- Have students investigate the meaning of *trade winds* and *horse latitudes* and explain how they are related to the Coriolis Effect.

Visit WorldRecordsBooks.com for more images and activities!

The World's Windiest Place
Activity Sheet 8A

Name _____ Class _____ Date _____

FOCUS *Read the entries from the* Meteorologist's Handbook *and from the* World Weather Records Book. *As you read, compare what you are reading about with any personal experiences you have had with this type of weather. Then answer the questions that follow.*

From the *Meteorologist's Handbook*
- **wind:** air moving through the atmosphere, especially across the Earth's surface
Wind speed is measured by a device called an *anemometer*.

From the *World Weather Records Book*
The **world's windiest place** is Port Martin, Antarctica, where the wind always blows hard, reaching speeds upwards of 200 mph.

1. Describe the location of the world's windiest place.

   _____
   _____
   _____

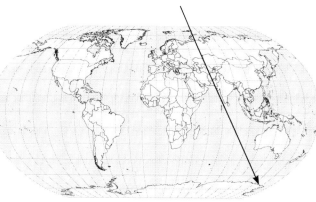

2. Port Martin is the world's windiest place, but it isn't the place where the world's highest wind speed was measured. That was at Mount Washington, in New Hampshire. What do you think makes Port Martin windier, overall, than Mount Washington?

   _____
   _____
   _____
   _____
   _____

Why Does the Wind Blow?

Activity Sheet 8B

Name _____ Class _____ Date _____

FOCUS *Why does the wind blow? Read the article. Then answer the questions.*

The Wind Blows From High to Low

The wind always blows for the same reason.

Because the Earth is a sphere, the Sun does not heat all areas of the Earth evenly. Where the Earth and the atmosphere receive more of the Sun's energy, the air tends to be warmer. Warmer air has a lower pressure than cooler air and tends to rise. When it rises, the cooler air around it, which is at a higher pressure, rushes in to replace it. This rushing of air is the wind.

At any given moment, there are countless low- and high-pressure areas in the atmosphere. Winds are constantly blowing from the high-pressure areas to the low-pressure areas, as the atmosphere naturally seeks to even-out the air pressure. As the Sun continues to pour its energy onto the Earth, more high- and low-pressure areas are continuously created, so the wind continues to blow.

When the difference between neighboring high- and low-air pressure is great, the wind blows harder. The wind also blows harder if these neighboring areas are closer together.

The combination of these two measures—difference in pressure and distance apart—is called the Pressure Gradient Force (PGF). The greater the PGF, the greater the wind speed as the air pressure tries to even-out.

Other factors, including the friction of the air blowing over the Earth's surface and the rotation of the Earth, also affect winds. But the main cause of winds is a difference in air pressure. This difference, caused by the uneven heating of the Sun, is the major cause of wind—and weather, generally.

1. Explain the title of the article.

   _____

2. What is the PGF?

   _____

3. What is the relationship between the PGF and wind speed?

   _____
   _____

Understanding the Coriolis Effect

Activity Sheet 8C

Name _____ Class _____ Date _____

FOCUS

"Winds blow from high to low." In other words, wind blows as air travels from high-pressure areas in the atmosphere to low-pressure areas. But the wind doesn't blow straight from one area to another. Wind, it might surprise you to learn, follows a curved path. Why? The answer is "because of the Coriolis Effect." Read about the Coriolis Effect. Then conduct the activity.

The Coriolis Effect

The Coriolis Effect, like many things involving the weather, is named for the scientist who first discovered or described it. Gustave-Gaspard Coriolis first described the effect that would be named for him in 1835.

The Coriolis Effect is the tendency of things moving across a spinning surface to curve. Since the Earth spins, things that move across it tend to curve. This includes air, which moves as wind. Winds follow a curving path across the Earth instead of blowing straight from high-pressure areas to low-pressure areas.

The Coriolis Effect also affects the weather in other ways. Ocean currents, which can have profound effects on weather, follow curved paths. The large circular storms of the oceans—hurricanes and typhoons—are affected by the Coriolis Effect.

Seeing the Coriolis Effect in Action

You can see how the Coriolis Effect works by performing the following activity:

1. Get a Lazy Susan. Cut a piece of paper the same size and shape of the top of the Lazy Susan and tape it in place.

2. Spin the Lazy Susan counterclockwise (the same direction the Earth rotates.)

3. Hold a ruler just above the spinning Lazy Susan, from the center to the edge. With a felt-tip or marking pen, draw a straight line on the paper along the ruler from the center to the edge.

4. Stop the Lazy Susan and look at the line. Even though you drew a straight line, the line is curved: a demonstration of the Coriolis Effect.

5. Answer this question: How is the Lazy Susan in this experiment like the Earth and how is the ink like the wind?

9. The World's Highest Wind Speed
231 mph

Objectives
- Identify and discuss the world's highest wind speed
- Analyze and explain the use of the Beaufort scale
- Construct and utilize a weather vane

Time and Special Materials
- About one class period
- Dowel, spool, scissors, glue, tape, clay, stones, cardboard

Teaching Tips
You may wish to introduce the topic by completing the first activity sheet as a class. The two additional activity sheets can be completed by students working independently, with partners, or in small groups.

Activity Sheet 9A
- Check student comprehension of the terms in the *Meteorologist's Handbook*; you may have students clip the entries from the *Meteorologist's Handbook* and the *World Weather Records Book* and add them to their compilations.
- Focus students' attention on the map.

Activity Sheet 9B
- Emphasize how the development of the Beaufort scale is evidence of the primary importance of weather in human life.
- Have students apply the Beaufort Scales frequently by observing conditions outside the classroom.

Activity Sheet 9C
- You may wish to have students cooperate to make a single classroom weather vane.
- Remind students that wind direction is recorded as the direction *from* which the wind blows.

Answers
- *Activity Sheet 9A* 1. 231 mph; 2. the White Mountains; 3. Answers will vary. Reward thoughtful responses.
- *Activity Sheet 9B* 1. hurricane force; 2. small tree branches move; 3. Have your community's average wind speed ready.
- *Activity Sheet 9C* 1. Assist students with construction of a weather vane. Reward honest effort.

Extension and Enrichment
- Challenge students to design an anemometer.
- Have students integrate their wind log into a climate profile of your community.

Visit WorldRecordsBooks.com for more images and activities!

The World's Highest Wind Speed
Activity Sheet 9A

Name _____ Class _____ Date _____

FOCUS *Read the entries from the* Meteorologist's Handbook *and from the* World Weather Records Book. *As you read, compare what you are reading with any personal experiences you have with this type of weather. Then answer the questions that follow.*

From the *Meteorologist's Handbook*

- **anemometer:** an instrument that measures wind speed
  Wind speed is expressed in miles per hour (mph).

- **weather vane:** an instrument that measures wind direction
  Wind direction is expressed in compass direction.

From the *World Weather Records Book*

The **world's highest surface wind speed** outside of a tornado was recorded on April 12, 1934, at the summit of Mount Washington, New Hampshire (elevation: 6,288 feet). The anemometer recorded a peak wind speed of 231 miles per hour.

1. What is the world's highest surface wind speed?

   _____

2. In which mountain range is Mount Washington located?

   _____

3. Describe the strongest wind you've ever experienced.

   _____
   _____
   _____
   _____

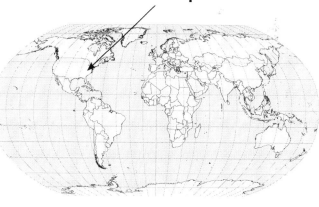

Mount Washington, New Hampshire

Wind and Its Effects
Activity Sheet 9B

Name _____ Class _____ Date _____

FOCUS

Sailors since ancient times have always been keenly interested in the winds. In 1806, an English admiral, Sir James Beaufort, devised a simple scale for sailors to use to measure wind. By looking at the way the sea was behaving and comparing it to the scale, sailors could judge the wind speed and rig their sails properly. This scale, called the Beaufort scale after its inventor, is still used today. Beaufort originally used numbers (0 through 12) and sea indicators (spray, foam, etc.) in his scale. Over the years, the scale was expanded so it could be used on land. Study the scale. Then answer the questions that follow.

Beaufort Number	Wind Speed (mph)	Sailor's Term	Effects at Sea	Effects on Land
0	Under 1	Calm	Sea like glass	Calm; smoke rises vertically
1	1-3	Light air	Ripples; no foam crests	Smoke drifts; vanes do not move
2	4-7	Light breeze	Small wavelets; crests not breaking	Wind felt on face; leaves rustle; vanes move
3	8-12	Gentle breeze	Large wavelets; crests breaking; scattered whitecaps	Leaves, twigs move; light flags extended
4	13-18	Moderate breeze	Small waves; numerous whitecaps	Dust, leaves, paper raised up; small branches move
5	19-24	Fresh breeze	Moderate waves; many whitecaps; some spray	Small trees sway
6	25-31	Strong breeze	Larger waves; whitecaps everywhere; more spray	Large branches in motion; whistling heard in wires
7	32-38	Moderate gale	Sea heaping up; white foam blowing in streaks	Whole trees in motion; resistance felt in walking against wind

Beaufort Number	Wind Speed (mph)	Sailor's Term	Effects at Sea	Effects on Land
8	39-46	Fresh gale	Moderately high waves; foam blowing in well-marked streaks	Twigs and small branches broken off trees
9	47-54	Strong gale	High waves, sea rolling; dense streaks of foam; spray may reduce visibility	Slight structural damage occurs; shingles blown from roofs
10	55-63	Whole gale	Very high waves with overhanging crests; sea looking white; visibility reduced	Seldom experienced on land; trees broken; structural damage occurs
11	64-72	Storm	Exceptionally high waves; sea covered with foam	Very rarely experienced on land; usually with widespread damage
12	73 or higher	Hurricane force	Air filled with foam; sea white with driving spray; visibility greatly reduced	Violence and destruction

1. The world's highest surface wind speed ever recorded was 231 miles per hour. What is the sailor's term for wind of this force?

2. How would you know by looking out a window at your school when a Beaufort number 4 wind was blowing?

3. Conduct research online to find out the average wind speed in your community. What is its Beaufort Number?

Constructing a Weather Vane
Activity Sheet 9C

Name _____ Class _____ Date _____

You can easily construct and use a weather vane to determine wind direction. And you can use the Beaufort scale to estimate wind speed. Just follow the steps below. Check off each one as you complete it.

FOCUS *Meteorologists are always eager to know two basic facts about any wind that blows: its speed and its direction. Wind speed is measured by an* **anemometer** *and expressed in miles per hour. Wind direction is measured by a* **weather vane** *and expressed by compass direction. Winds are identified according to the direction* **from** *which they blow, not the direction they are blowing toward.*

____ **Step 1 Construct a Weather Vane.**
Follow the illustration. Your teacher will help you.

____ **Step 2 Set up Your Weather Vane.**
Your teacher will guide you.

____ **Step 3 Keep a Wind Log.**
Note the wind direction and estimated speed each day, and record them in a wind log.

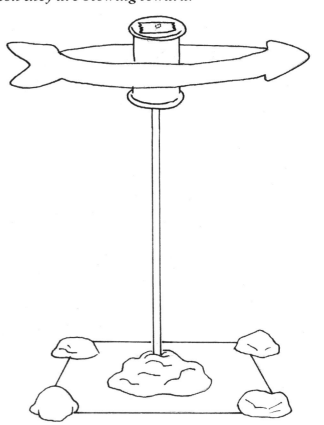

Date	Wind Direction	Beaufort Number; Estimated Wind Speed
April 7	West	2; 4-7 mph
April 8	Northwest	3; 8-12 mph

10. The World's Wettest Place
Mawsynram, India

Objectives
- Identify and discuss the world's wettest place
- Define *monsoon*
- Explain why it rains

Time and Special Materials
- About one class period
- Spray bottle filled with water
- Mirror or window

Teaching Tips
You may wish to introduce the topic by completing the first activity sheet as a class. The two additional activity sheets can be completed by students working independently, with partners, or in small groups.

Activity Sheet 10A
- Check student comprehension of the term in the *Meteorologist's Handbook*; you may have students clip the entries from the *Meteorologist's Handbook* and the *World Weather Records Book* and add them to their compilations.
- Focus students' attention on the map.

Activity Sheet 10B
- Emphasize that a monsoon is actually a wind pattern and not a rain.
- Have students identify monsoon regions on a world map.

Activity Sheet 10C
- Encourage students to perform the spray bottle experiment.
- Point out that meteorologists still don't know what causes rain to fall in all situations.

Answers
- *Activity Sheet 10A* 1. Asia, Southern Asia, India; 2. 1.25 inches.
- *Activity Sheet 10B* 1. A strong wind from the sea, caused by the rising of warm air over the water; 2. Answers will vary, but should include that strong rains coincide with monsoons; 3. Monsoons are seasonal.
- *Activity Sheet 10C* 1. Water on the Earth's surface evaporates. In the atmosphere, the water vapor condenses into very small water droplets and forms clouds; 2. The water droplets that make up clouds are extremely small and light. They fall at a rate of just 2 feet a minute, and the smallest updraft will send them up; 3. Water droplets stick together when they touch and fall when they are large enough.

Extension and Enrichment
- Challenge students to name as many forms of precipitation as they can.
- Locate and display pictures of the awesome monsoon rains in India.

Visit WorldRecordsBooks.com for more images and activities!

The World's Wettest Place
Activity Sheet 10A

Name _____ Class _____ Date _____

FOCUS *Read the entries from the* Meteorologist's Handbook *and from the* World Weather Records Book. *As you read, compare what you are reading with any personal experiences you have had with this type of weather. Then complete the activities that follow.*

From the *Meteorologist's Handbook*
- **rain:** drops of water that condense from moisture in the atmosphere and fall to Earth

From the *World Weather Records Book*
The **world's wettest place** is Mawsynram, India. On average, Mawsynram receives about 467 inches of rain per year. That's more than 10 times as much rain as most cities in the United States receive.

1. Describe the location of the world's wettest place.

   _____
   _____
   _____
   _____
   _____

2. If the rain in Mawsynram fell evenly during the year, how much rain would fall each day?

   _____

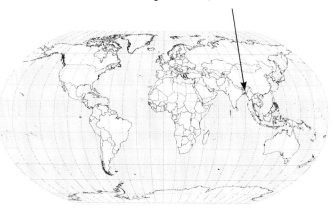

Mawsynram, India

54 *Amazing World Records • Weather*   SunflowerEducation.net

Understanding Monsoons
Activity Sheet 10B

Name _____ Class _____ Date _____

FOCUS *The world's wettest place is wet because of* **monsoons**. *But a monsoon is not a rain. To find out what one really is, read the article. Then answer the questions.*

Monsoons

The world's wettest place is Mawsynram, India, which receives an average of 467 inches of rain each year. Rain doesn't fall there every day, though. In Mawsynram, most of the rain falls from April to October, during the monsoon season.

What's a monsoon? People often call great rainstorms monsoons, but a monsoon isn't a rain. It's really a wind.

There are monsoons all around the world, but the most famous are those of the Indian Ocean. There, as elsewhere, the monsoon is created by the different temperatures of the land and the sea.

During the summer, the land heats up more than the water. This causes higher temperatures and lower air pressure in the atmosphere over the land. This hot air rises and is replaced by the cooler, lower-pressure air that rushes in from over the ocean. This air, which carries a great deal of moisture from the ocean, blows in, bringing a great amount of rainfall.

In the winter, the process is reversed. The winds blow out to sea, and it very rarely rains. So climates in monsoon regions are characterized by definite wet and dry seasons. Often, the wet season is called the monsoon season, and the rains are called monsoons. But remember, a monsoon is actually a wind.

1. Write a definition of *monsoon*.

   _____

   _____

2. Why do you think people sometimes call any heavy rainfall a monsoon?

   _____

   _____

3. The word monsoon comes from an Arabic word that means "season." Why is this appropriate?

   _____

What Makes Rain?
Activity Sheet 10C

Name _____ Class _____ Date _____

FOCUS *Why does it rain? This question has intrigued people for tens of thousands of years. Only recently, however, have scientists developed a firm understanding of the actual processes that make rain. But here's something that may surprise you: We still don't know in every instance what causes rain to fall. To find out what we do know, read the article. Then answer the questions.*

Why Is Your Picnic Ruined?

You know that rain falls from clouds. But why is it that many clouds don't produce rain?

You also know that clouds contain water vapor. But how can clouds made of water float in the sky?

To answer these mysteries, you need to understand where clouds come from. The atmosphere contains water in the form of vapor, a gas. This vapor forms when liquid water on the Earth's surface—oceans, lakes, puddles, and so on—is heated by the Sun and evaporates. When this moist air rises, it cools. The water vapor in the air then condenses, or turns back into liquid, and forms clouds.

But why doesn't the water immediately fall back to Earth as rain? The answer is that the water vapor condenses into very, very small droplets. In fact, the droplets in clouds are so small that they're only about one one-millionth the size of a typical raindrop. Because they're so very, very small and light, they fall very, very slowly—less than two feet in a minute. The slightest, lightest upward movement of air pushes them back up.

It's like when you blow soap bubbles. They can fly up and away from you; but, in perfectly still air, they fall. And cloud droplets are much smaller and lighter than soap bubbles.

But if the air were perfectly still, these droplets still wouldn't fall all the way to the ground. Why? Again, because of their size. You know how smaller puddles dry up faster than bigger puddles after a rain? Cloud droplets are so small that they'd dry up—evaporate back into vapor—long before they fell to the ground.

So what makes the cloud droplets turn into rain? There are two answers. One, called *coalescence*, explains the process in this way: The larger cloud droplets begin to fall and run into other droplets. As they combine, they get heavier, bigger, and fall faster. A droplet falling a mile would combine with about one million other cloud droplets, forming a raindrop heavy enough to drop to Earth.

Another explanation is called *ice crystal formation*. When the air is cold enough, ice crystals form in clouds. These grow large and heavy enough to begin falling and combine with other ice crystals as they fall. Scientists believe these crystals melt on the way down, forming raindrops.

Both the coalescence and ice crystal explanations of rain creation are true but incomplete. So scientists are still looking for the answer to the age-old question: What makes rain?

1. Why do clouds form?

2. Why doesn't the water in clouds immediately fall back to Earth as rain?

3. Use a spray bottle to spray more and more droplets of water onto a mirror or a window. Watch what happens. How does this demonstrate the coalescence explanation of rain formation?

11. The World's Greatest Rainfall
73.62 inches

Objectives
- Identify and discuss the world's greatest rainfall
- Construct and utilize a rain gauge
- Differentiate among different classifications of liquid precipitation

Time and Special Materials
- About one class period
- Coffee can, ruler, tape

Teaching Tips
You may wish to introduce the topic by completing the first activity sheet as a class. The two additional activity sheets can be completed by students working independently, with partners, or in small groups.

Activity Sheet 11A
- Check student comprehension of the terms in the *Meteorologist's Handbook*; you may have students clip the entries from the *Meteorologist's Handbook* and the *World Weather Records Book* and add them to their compilations.
- Focus students' attention on the map.

Activity Sheet 11B
- Have students mount their rain gauge to National Weather Service specifications, with its rim 12 inches about the surface of the ground.
- Rotate rain gauge emptying and recording responsibilities among the students in the class.

Activity Sheet 11C
- Have students listen for these specific terms in local weather forecasts.
- Challenge students to correctly classify local rainfalls according to the chart.

Answers
- **Activity Sheet 11A** 1. 3 inches per hour; 2. Answers will vary. Reward thoughtful and creative responses.
- **Activity Sheet 11B** Assist students with creating a rain gauge and a rainfall log. Reward honest effort and diligence in taking logs.
- **Activity Sheet 11C** 1. drizzle and rain; 2. Rain is made up of drops greater than .02 inches in diameter. Drizzle is made up of drops smaller than .02 inches in diameter; 3. distance of visibility; 4. rate at which rain falls; 5. Answers will vary. Reward thoughtful responses; 6. Answers will vary and may include the following responses: meteorologists and pilots. Reward thoughtful responses.

Extension and Enrichment
- Invite a representative for the local office of the National Weather Service to make a presentation about precipitation and answer students' questions.
- Have students integrate their rain log into a climate profile of your community.

Visit WorldRecordsBooks.com for more images and activities!

The World's Greatest Rainfall
Activity Sheet 11A

Name _____ Class _____ Date _____

FOCUS *Read the entries from the* Meteorologist's Handbook *and from the* World Weather Records Book. *As you read, compare what you are reading with any personal experiences you have had with this type of weather. Then answer the questions that follow.*

From the *Meteorologist's Handbook*
- **rain gauge:** an instrument used to measure rainfall
  Rainfall is expressed in inches.

From the *World Weather Records Book*
The **world's greatest rainfall** ever recorded over a 24-hour period was 73.62 inches at Cilaos on Reunion Island in the Indian Ocean, on March 15 and 16, 1952.

1. During the record-setting 24 hour period, how much rain fell at Cilaos, on average, each hour?

   _____
   _____

2. Describe the heaviest rainfall you've ever experienced.

   _____
   _____
   _____
   _____
   _____
   _____
   _____

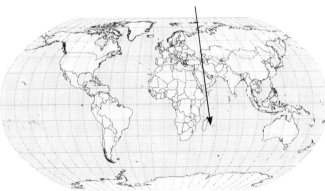

Reunion Island

Constructing a Rain Gauge
Activity Sheet 11B

Name _____ Class _____ Date _____

FOCUS *Rainfall is measured in inches. Meteorologists use devices called **rain gauges** to measure rainfall. Some rain gauges measure rainfall by how much rain fills up a container. Other rain gauges weigh the amount of rain in a container to determine rainfall. Some rain gauges are very sophisticated and are connected to computers, and some even empty themselves. But you don't need such a sophisticated device to measure rainfall. In fact, you can make a simple rain gauge yourself. Just complete the steps below. Check off each one as you complete it.*

____ **Step 1** **Construct a Rain Gauge.** Follow the illustration. Your teacher will help you.

- 2-liter soda bottle
- 1-inch hash marks on side

____ **Step 2** **Set up Your Rain Gauge.** Your teacher will guide you.

____ **Step 3** **Keep a Rainfall Log.** Note the amount of rainfall after each rainstorm. Record your findings in a log. Your log should look something like this:

Date	Rainfall
April 27	1/4 inch
April 28	none
April 29	1.2 inch

Classifying Rain
Activity Sheet 11C

Name _____ Class _____ Date _____

FOCUS *You know that rain is liquid precipitation. You also know, firsthand, that there are different kinds of rain—sprinkles, downpours, even "raining cats and dogs." But did you know that meteorologists have official definitions for different kinds of liquid precipitation? They're explained in the chart below. Study the chart. Then use it to answer the questions.*

Liquid Precipitation	
Drizzle *Drizzle* is liquid precipitation with water drops less than .02 inches in diameter that fall close together	• *Light Drizzle* is drizzle in which visibility is greater than 10/16 of a mile • *Moderate Drizzle* is drizzle in which visibility is between 5/16 and 10/16 of a mile • *Heavy Drizzle* is drizzle in which visibility is less than 5/16 of a mile
Rain *Rain* is liquid precipitation with water drops greater than .02 inches in diameter that fall far apart or with smaller water drops that fall far apart	• *Light Rain* is rain that falls at a rate of .10 inches or less an hour • *Moderate Rain* is rain that falls at a rate of between .11 and .30 inches an hour • *Heavy Rain* is rain that falls at a rate of more than .30 inches an hour

1. What are the two forms of liquid precipitation?

   _____

2. What is the difference between drizzle and rain?

   _____
   _____
   _____
   _____

3. What measure is used to separate the different types of drizzle?

_____
_____
_____

4. What measure is used to separate the different types of rain?

_____
_____
_____

5. Why do you think the different types of rainfall and drizzle are separated?

_____
_____
_____

6. Who might need to know specifically what type of rain or drizzle is falling?

_____
_____
_____

12. The World's Greatest Snowfall
1,224 1/4 inches

Objectives
- Identify and discuss the world's greatest snowfall
- Explain why and how snow forms and falls
- Investigate how snow affects human activities

Time Required
- About one class period

Teaching Tips
You may wish to introduce the topic by completing the first activity sheet as a class. The two additional activity sheets can be completed by students working independently, with partners, or in small groups.

Activity Sheet 12A
- Check student comprehension of the term in the *Meteorologist's Handbook*; you may have students clip the entries from the *Meteorologist's Handbook* and the *World Weather Records Book* and add them to their compilations.
- Focus students' attention on the map.

Activity Sheet 12B
- Make sure students understand the concept of supercooled water.
- Ensure students understand the relationship between snow or ice crystals and snowflakes.

Activity Sheet 12C
- This activity works well with students paired as partners.
- Encourage students to expand their answers beyond firsthand experiences.

Answers
- **Activity Sheet 12A** 1. 3 1/3 inches; 2. Answers will vary. Reward thoughtful responses; 3. Answers will vary. Reward thoughtful responses.
- **Activity Sheet 12B** 1. Water droplets freeze and crystallize. The ice crystals grow; and as they fall, they join with other ice crystals to form a snowflake; 2. Snowflakes are made of ice crystals; 3. the temperature at which they form.
- **Activity Sheet 12C** Answers will vary. Reward thoughtful and creative responses.

Extension and Enrichment
- You may wish to challenge students to find the answer to the questions, "Are two snowflakes ever exactly alike?"
- Challenge students to explain why accurately measuring snowfall is more difficult than accurately measuring rainfall.

Visit WorldRecordsBooks.com for more images and activities!

The World's Greatest Snowfall
Activity Sheet 12A

Name _____ Class _____ Date _____

FOCUS *Read the entries from the* Meteorologist's Handbook *and from the* World Weather Records Book. *As you read, compare what you are reading with any personal experiences you have had with this type of weather. Then answer the questions that follow.*

From the *Meteorologist's Handbook*
- **snow:** particles of water vapor that freeze in the atmosphere and fall to Earth as crystal flakes

From the *World Weather Records Book*
The **world's greatest snowfall** in any given year was recorded at Paradise Ranger Station, Mount Rainier, Washington, from February 1971 to February 1972. A total of 1,224 ¼ inches of snow fell.

1. On average, how much snow fell each day in Paradise from February 1971 to February 1972?
   _____

2. Does Paradise seem like "paradise" to you? Explain why or why not you'd like to experience so much snowfall.

   Mount Rainier, Washington

   _____
   _____
   _____
   _____
   _____

3. What do you like best about snow?
   _____
   _____

Why Does It Snow?
Activity Sheet 12B

Name _____ Class _____ Date _____

FOCUS *Snow is precipitation in the form of ice crystals. But why does it snow? It's a good question. Why do you think it snows? To see if you're right, read the following article. Then answer the questions that follow.*

Why It Snows

Water evaporates from the ground and condenses into the tiny water droplets that form clouds. These droplets can turn into raindrops or hail or return to water vapor. If the conditions are right, however, they can turn into snowflakes.

If a droplet of water in a cloud gets cold enough, it will freeze into an ice crystal. At the right temperature and with enough water vapor in a cloud, this crystal will grow. As more water vapor condenses onto it, the crystal grows larger and larger.

Water droplets that are *supercooled*—that is, colder than freezing but still in liquid form—may also join the growing crystal, freezing when they contact it. The crystal continues to grow heavier, until it begins to drop from the sky. As it drops, it joins with other crystals, making a snowflake.

A very large snowflake—say, an inch across—may be made up of about 100 crystals. Crystals grow into a variety of shapes. The shape of the crystals, and the flakes they form, is determined by the temperature at which they form.

There's a lot of air in between the snowflakes in a handful of snow. That's why you can compact it to form a snowball or a snowman. That's also why snow on the ground takes up a lot more room than rain. On average, a ten-inch snowfall would melt down to the equivalent of about an inch of rain.

1. List the major steps in the formation of a snowflake.

   _____

2. What is the relationship between ice crystals and snowflakes?

   _____

   _____

3. What determines the shape of ice crystals and snowflakes?

   _____

Snow
Activity Sheet 12C

Name _____ Class _____ Date _____

FOCUS *Think about your experiences and the experiences of other people in snow. Use what you know about snow to complete the chart.*

People and Snow	
Ways People Travel on Snow	
Problems for People Caused by Snow	
How People Make Money from Snow	
How Snow Affects Peoples' Clothing and Shelter	
Your Best Snow Experience	

13. The World's Driest Place
The Coast of Chile

Objectives
- Identify and discuss the world's driest place
- Explore human attempts at weather modification
- Define and discuss the concept of drought

Time Required
- About one class period

Teaching Tips
You may wish to introduce the topic by completing the first activity sheet as a class. The two additional activity sheets can be completed by students working independently, with partners, or in small groups.

Activity Sheet 13A
- Check student comprehension of the terms in the *Meteorologist's Handbook*; you may have students clip the entries from the *Meteorologist's Handbook* and the *World Weather Records Book* and add them to their compilations.
- Focus students' attention on the map.

Activity Sheet 13B
- Point out that human attempts to make it rain are age-old.
- Guide a class debate on the pros and cons of human attempts to modify natural processes.

Activity Sheet 13C
- Make sure students understand the devastating consequences of drought in regions of subsistence farming.
- Help them see how droughts have serious, though less dire, consequences elsewhere as well.

Answers
- **Activity Sheet 13A** 1. about 500 miles; 2. Answers will vary and may include the following responses: sparse; tough; can store lots of water. Reward thoughtful responses.
- **Activity Sheet 13B** 1. appealing to gods and goddesses, firing cannons at clouds, and lofting explosive charges in the air with balloons; 2. dropping dry ice or silver iodide into clouds to give the water droplets something on which to form.
- **Activity Sheet 13C** 1. a period of unusually little rainfall; 2. The landscape changes as plants suffer and water levels drop; 3. People who depend on the food they grow to eat lose their food source.

Extension and Enrichment
- Have students research and report on the Dust Bowl of the 1930s.
- Have students research and report on irrigation techniques.

Visit WorldRecordsBooks.com for more images and activities!

The World's Driest Place
Activity Sheet 13A

Name _____ Class _____ Date _____

FOCUS *Read the entries from the Meteorologist's Handbook and from the World Weather Records Book. As you read, compare what you are reading with any personal experiences you have had with this type of weather. Then answer the questions that follow.*

From the *Meteorologist's Handbook*
- **desert:** an area that receives less than 10 inches of precipitation per year

From the *World Weather Records Book*
The **world's driest place** is along the Pacific Coast of Chile in South America, between Arica and Antofagasta, where there is virtually no rain. Since records have been kept, the average annual rainfall has been less than .004 inches. No rain fell at all in Arica for one 14-year period.

1. What would you expect human, animal, and plant life in this region to be like?

   _____

   _____

2. People often describe this region as "a moonscape." What do you think they mean? How is this description related to the fact that it is the world's driest place?

The Coast of Chile

   _____

   _____

   _____

   _____

   _____

   _____

68 Amazing World Records • Weather

Rainmaking
Activity Sheet 13B

Name _____ Class _____ Date _____

FOCUS — *Mark Twain is often quoted as saying, "Everybody talks about the weather, but nobody does anything about it." It's a funny thing to say, but it isn't true at all. To find out what people have done, or at least tried to do, read on. Then answer the questions.*

Since ancient times, people have tried to make it rain. Even the earliest peoples knew that rain was needed for plants to grow and for animals and people to live. They knew that rain meant life, and they did their best to make sure there was enough of it. Many early peoples appealed to gods and goddesses to bring the rain.

People have also tried to stop the rain. In medieval Europe, church bells were rung in an attempt to ward off thunderstorms that would bring crop-damaging hail, frightening thunder, too-heavy rainfall, and fire-starting lightning. Ironically, many bell-ringers were killed by lightning from the very storms they tried to prevent.

By the 1800's, some people thought that loud noises—a kind of artificial thunder—could set off rainstorms. So they fired cannons toward the clouds. They lofted explosive charges into the air with balloons and kites. Not surprisingly, none of these things worked.

Today, people are still trying to make it rain—and sometimes do. Since people now have a scientific understanding of what causes rain, they have more success as rainmakers. The most successful attempts result from *cloud seeding*. In cloud seeding, an airplane drops dry ice or a chemical called silver iodide into the clouds. In the clouds, both chemicals produce ice crystals that are heavy enough to fall and to start forming snow or rain. Cloud seeding, however, isn't usually very successful. Mostly, it is used in an attempt to increase the amount of precipitation that is going to fall anyway. Still, it's a big improvement over shooting cannons into the sky.

1. Describe some early attempts to make it rain.

_____

_____

2. What is cloud seeding?

_____

_____

Understanding Drought
Activity Sheet 13C

Name _____ Class _____ Date _____

FOCUS *You may have heard of the term* **drought**, *but do you know exactly what it is? Do you know why droughts can be so devastating? Read the Fact Box to learn the answers to these questions. Then answer the questions at the bottom of the page.*

Fact Box: Droughts

- The word drought comes from an old word that means "dry." The preferred way to pronounce it is "drout."

- Droughts occur in areas that receive relatively substantial rainfall. When the area is abnormally dry for an extended period of time, we say that a drought is occurring.

- A drought is defined by its unusualness. There really aren't droughts in a desert, for example, because it usually doesn't rain much anyway.

- There is no "official" definition for how much less rain has to fall or for how long the dry period continues before people call it a drought. Usually, though, the word drought is used when the dry period causes the landscape to change as plants (especially crops) suffer, the level of streams, ponds, and lakes drop, and so on.

- Droughts are devastating for people who live by subsistence farming. Without enough rain, the crops fail to grow, or die, and there isn't enough food. Famine can result. Areas in Africa, where long droughts have resulted in famines that have killed millions of people, have especially suffered in recent times.

1. Write a definition of the word *drought*.

   _____

2. What happens to the landscape during a drought?

   _____
   _____

3. Why can droughts be so deadly?

   _____
   _____
   _____

14. The World's Foggiest Place
The Grand Banks

Objectives
- Identify and discuss the world's foggiest place
- Explain how different types of fog form
- Explore how sailors cope with fog at sea

Time Required
- About one class period

Teaching Tips
You may wish to introduce the topic by completing the first activity sheet as a class. The two additional activity sheets can be completed by students working independently, with partners, or in small groups.

Activity Sheet 14A
- Check student comprehension of the term in the *Meteorologist's Handbook*; you may have students clip the entries from the *Meteorologist's Handbook* and the *World Weather Records Book* and add them to their compilations.
- Focus students' attention on the map.

Activity Sheet 14B
- Stress the similarity between fog and clouds in the sky.
- Emphasize the process through which fog forms.

Activity Sheet 14C
- If you live in a coastal area, encourage students to identify the sounds made by ships in the fog.
- Give students the opportunity to create these sounds and challenge their classmates to correctly identify them.

Answers
- *Activity Sheet 14A* 1. three; 2. Answers will vary. Reward thoughtful responses.
- *Activity Sheet 14B* 1. Fog is made of tiny droplets of water; 2. Warm air that has reached the holding capacity of water vapor cools. At the dew point, the water vapor condenses and forms fog; 3. Check tables to ensure type of fog, how fog is formed, and where fog is formed are all addressed.
- *Activity Sheet 14C* 1. Sailors rely very heavily on their sight to be able to steer clear of obstacles and other ships; 2. Because sound travels much farther in fog than light does; 3. Answers will vary. Reward thoughtful responses.

Extension and Enrichment
- Fog has intrigued people for thousands of years. Have students write poems or paragraphs that express the mysterious qualities of fog.
- Challenge students to identify the types of fog that appear in or near your community.

Visit WorldRecordsBooks.com for more images and activities!

The World's Foggiest Place
Activity Sheet 14A

Name _____ Class _____ Date _____

FOCUS *Read the entries from the* Meteorologist's Handbook *and from the* World Weather Records Book. *As you read, compare what you are reading with any personal experiences you have had with this type of weather. Then answer the questions that follow.*

From the *Meteorologist's Handbook*
- **fog:** a cloud that forms on the surface of the Earth

From the *World Weather Records Book*
The **world's foggiest place** is on the Grand Banks, off the coast of Newfoundland, Canada, in the North Atlantic Ocean. There, fog lasts for weeks at a time with an average of 120 days of dense fog per year.

1. Fill in the blank: The Grand Banks is covered in fog one out of about every _____ days a year.

2. How do you think fog affects human activity on the Grand Banks?

_____

_____

_____

_____

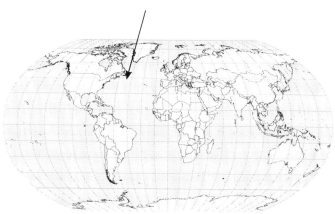

The Grand Banks

What Makes Fog?
Activity Sheet 14B

Name _____ Class _____ Date _____

FOCUS *Fog is perhaps the spookiest weather phenomenon. To take the mystery out of fog, read the article. Then answer the questions.*

Fog

If you want to know what it's like inside a cloud, just go for a walk in the fog. Fog is a cloud that forms on the ground.

Like clouds high in the sky, fog is made of tiny droplets of water. This water comes from water on the surface—oceans, rivers, and so on—that has evaporated near the ground into water vapor.

But the air can only hold so much water vapor. This is called the *holding capacity* of the air. When the temperature of the air decreases, its holding capacity decreases, too. When the air is cool enough, the water vapor that it is holding condenses, or turns back into liquid water. This water takes the form of billions of tiny water droplets. When this happens, fog is formed. The temperature at which the air reaches its holding capacity is called the *dew point*.

All fog is formed in this basic way. But, depending on the circumstances that cause this process, fog is classified into different types.

Advection fog forms when wind blows water vapor-filled air from a warm area over a cooler area. As the air reaches the cooler area, its temperature drops, the vapor begins to condense, and fog is formed. The kind of fog along seacoasts is usually advection fog, as the warm, moist air over the sea blows onshore over the cooler land. It's often called sea fog.

You've probably seen *radiation fog* on cool mornings. It occurs on clear nights with little wind, when the ground cools during the night. A layer of fog, usually just 100 feet deep or so, forms in this cooler area next to the ground. As the day heats up, the capacity of the air to hold water increases, and the fog "burns off."

Upslope fog occurs, as the name suggests, when wind blows up slopes. As the air rises, it cools and forms fog as its holding capacity decreases. *Frontal fog* forms along a front—the boundary between two air masses of different temperatures.

The boundaries of fronts are not straight up and down, but instead, they're at an angle—with the colder air mass, which is heavier, pushing under the warmer air mass, which is lighter. As rain from the warmer air drops into the colder air, the colder air reaches its holding capacity, and fog is formed.

All of these types of fog are caused by warm, moist air being cooled. But so-called *steam fog* or *sea smoke fog* is caused by cold air absorbing moisture from warming water. Water evaporates into this cold air, quickly saturating it to the dew point, and condenses into fog. This is the "steam" you see rising from ponds and streams early in the morning.

1. How is fog like a cloud?

   _____
   _____
   _____

2. Describe how fog forms using the terms *holding capacity* and *dew point*.

   _____
   _____
   _____
   _____
   _____

3. Complete the table below.

Type of Fog	How Formed	Where Formed

Ships in Fog
Activity Sheet 14C

Name _____ Class _____ Date _____

FOCUS *Fog is of vital concern to the sailors aboard the hundreds of thousands of ships that must sometimes sail in it. In fact, many people associate fog with the low, haunting sounds of the foghorns of ships. The chart below explains what exactly these spooky noises mean. Use the chart to answer the questions.*

Sound Signals Ships Use in Fog

Message	Foghorn Signal	Bell and Gong Signal
under sail	one long blast and two short blasts—every two minutes	
sailboat under power	one long blast—every two minutes	
under way	two long blasts—every two minutes	
aground (small ship)		three bells, followed by a rapid ringing, followed by another three bells—every minute
aground (large ship)		three bells, followed by a rapid ringing, followed by another three bells, followed by a gong—every minute
at anchor (small ship)		rapid ringing—every minute
at anchor (large ship)		rapid ringing followed by a gong—every minute

1. A sailor once wrote that "The skipper who sets out in a fog is courting danger." Why is this so?

_____

_____

_____

2. Why must sound signals be used in fog?

_____

_____

_____

_____

3. What do you think is the greatest danger of sailing in fog?

_____

_____

_____

_____

_____

15. The World's Heaviest Hail
Fell in Bangladesh

Objectives
- Identify and discuss the world's heaviest hailstones
- Explain how and why hail forms
- Investigate freak hail

Time Required
- About one class period

Teaching Tips
You may wish to introduce the topic by completing the first activity sheet as a class. The two additional activity sheets can be completed by students working independently, with partners, or in small groups.

Activity Sheet 15A
- Check student comprehension of the terms in the *Meteorologist's Handbook*; you may have students clip the entries from the *Meteorologist's Handbook* and the *World Weather Records Book* and add them to their compilations.
- Focus students' attention on the map.

Activity Sheet 15B
- Make sure students can differentiate between hail and hailstones.
- Explain how damaging hail can be to such things as automobiles, and more seriously, to crops.

Activity Sheet 15C
- Emphasize the extreme rarity of such occurrences.
- Point out that these are documented events.

Answers
- *Activity Sheet 15A* 1. At least one-fifth inch in diameter. Check circles to ensure they are of one-fifth inch in diameter; 2. Check circles to ensure they are softball sized (about 3.8 inches in diameter).
- *Activity Sheet 15B* 1. There must be supercooled water droplets and frozen raindrops in the upper atmosphere; 2. a thunderstorm; 3. Hail can destroy crops and property, and hurt or even kill people.
- *Activity Sheet 15C* 1. because they are very unusual; 2. normal sized hailstones combining during the right—and extremely rare—conditions; 3. Answers will vary. Reward thoughtful responses.

Extension and Enrichment
- Remind students that hail can be extremely dangerous and they should always seek shelter from it immediately.
- Have students take note of the idiosyncratic and sometimes amusing way local television weather forecasters classify hail (pea-size, turtle-head-size, etc.)

Visit WorldRecordsBooks.com for more images and activities!

The World's Heaviest Hail
Activity Sheet 15A

Name _____ Class _____ Date _____

FOCUS *Read the entries from the Meteorologist's Handbook and from the World Weather Records Book. As you read, compare what you are reading about with any personal experiences you have had with this type of weather. Then complete the activities that follow.*

From the *Meteorologist's Handbook*
- **hail:** balls of ice at least one-fifth inch in diameter that sometimes fall during thunderstorms

From the *World Weather Records Book*
The **world's heaviest hailstones** fell in Bangladesh on April 14, 1986. They weighed about two and a quarter pounds each and killed dozens of people.

1. How large must a ball of ice be to be classified as hail? Draw a circle this size.

   _____

   _____

2. The record-setting hailstones that fell in Bangladesh were about the size of softballs. Draw a circle on the back of this sheet to indicate the approximate diameter of this hail.

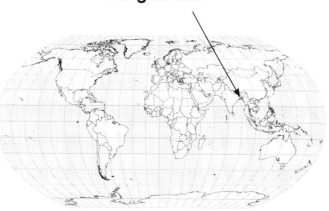

Bangladesh

What Makes Hail?

Activity Sheet 15B

Name _____ Class _____ Date _____

FOCUS *Read the facts about hail to learn how hail forms. Then answer the questions.*

Fact Box: Hail

- The word *hail* refers to a collection of individual *hailstones*.
- Hail is a form of frozen precipitation.
- Hail typically forms in thunderstorms.
- Hailstones form when droplets of supercooled water (water that is below freezing but still in liquid form) strike a frozen raindrop. The frozen raindrop has not yet fallen because it is small and light and the strong updrafts of the storm keep it aloft. As the supercooled water strikes the frozen raindrop, it freezes onto it, causing the hailstone to grow. When the hailstone grows too heavy for the updrafts to keep it aloft, it falls to the ground.
- The size of hailstones is largely determined by the intensity of the thunderstorm's updrafts. Faster updrafts create larger hailstones.
- A large hailstone will fall at a speed faster than 100 miles per hour.
- Hail usually falls for a short period of time, but some hailstorms last more than an hour.
- Hail can destroy crops, break windows, damage cars, and cause serious injury and even death.

1. What condition must exist for a hailstone to fall from the sky?

   _____
   _____

2. What weather condition typically accompanies hail?

   _____
   _____

3. Why can hail be a very serious weather condition?

   _____
   _____

Huge Hailstones
Activity Sheet 15C

Name _____ Class _____ Date _____

FOCUS *Most hailstones that fall are tiny—smaller than a pea. However, hail can grow much larger—even larger than a softball. But a few hailstones that have fallen from the sky are absolutely gigantic. Read about them, and answer the questions.*

Fact Box: Freak Hailstones

- One of the largest hailstones to fall in the United States was found in Coffeyville, Kansas, in 1970. It weighed a pound and a half and measured 17 inches around.
- Nearly a century earlier, in 1882, a huge hailstone that weighed 80 pounds fell on the Kansas farmland.
- In 1979, a large hailstone fell and killed a baby in Colorado.
- A 14-pound hailstone fell in England in 1972.
- Reports from India, where severe hailstorms are relatively common, put the weight of a few freak hailstones in excess of 100 pounds.
- Also in India, 200 years ago, people spoke of a hailstone that was the size of an elephant and that took days to melt.
- Freak hailstones are most likely caused by normal-sized hailstones combining during the right—and extremely rare—conditions.

1. Why are these hailstones called "freak" hailstones?

   _____

   _____

2. What probably causes freak hailstones?

   _____

   _____

3. Read the second-to-last statement in the Fact Box. Do you believe it? Why or why not?

   _____

   _____

World Records of Weather Disasters

16. The World's Deadliest Hurricane
The Galveston Hurricane of 1900

Objectives
- Identify and discuss the world's deadliest hurricane
- Investigate the formation, structure, and other important aspects of hurricanes
- Track a hurricane

Time Required
- About one class period

Teaching Tips
You may wish to introduce the topic by completing the first activity sheet as a class. The two additional activity sheets can be completed by students working independently, with partners, or in small groups.

Activity Sheet 16A
- Check student comprehension of the terms in the *Meteorologist's Handbook*; you may have students clip the entries from the *Meteorologist's Handbook* and the *World Weather Records Book* and add them to their compilations.
- Focus students' attention on the map.

Activity Sheet 16B
- Remind students that hurricanes, cyclones, and typhoons are essentially the same types of storms that occur in different parts of the world.
- Explain that the storm surge, which causes coastal flooding, is the facet of hurricanes that causes the most damage and loss of life.

Activity Sheet 16C
- You may wish to perform this activity as a class.
- The locations of hurricanes are readily available from various websites and from NOAA Weather Radio.

Answers
- *Activity Sheet 16A* 1. We are unsure how many people were killed; 2. greater than 74 miles per hour; 3. Answers will vary. Reward thoughtful responses.
- *Activity Sheet 16B* 1. Around North America, in the Indian Ocean and the west Pacific Ocean; there are the right water, wind, and temperature conditions; 2. a huge amount of water pushed inland by the hurricane; 3. from the West Indian word *huran* meaning "big wind"; 4. Answers will vary. Reward thoughtful responses; 5. Hurricanes rotate around an area of low pressure. Very warm water and air are required to form a hurricane.
- *Activity Sheet 16C* Track a current or historical storm.

Extension and Enrichment
- Explain what sea walls are and how they have significantly decreased the damage from hurricane storm surges.
- Have students prepare reports on individual hurricanes.

Visit WorldRecordsBooks.com for more images and activities!

The World's Deadliest Hurricane
Activity Sheet 16A

Name _____ Class _____ Date _____

FOCUS *Read the entries from the* Meteorologist's Handbook *and from the* World Weather Records Book. *As you read, compare what you are learning with any personal experiences you have had with this type of weather. Then complete the activities that follow.*

From the *Meteorologist's Handbook*

- **hurricane:** a powerful circular storm, about 200 to 300 miles across, that develops in tropical areas and produces winds of greater than 74 miles per hour.
  Circular storms that form in the Indian Ocean are called *cyclones*; such storms that form in the western Pacific Ocean are called *typhoons*.

From the *World Weather Records Book*

The **world's deadliest hurricane** was the Galveston hurricane of 1900. At least 6,000 people, and as many as 12,000, were killed.

1. The World Records Weather Book lists two figures for the number of people killed by the Galveston Hurricane. Why do you think this is so?

   _____

   _____

2. How high do the wind speeds of a circular storm have to be to be classified as a hurricane?

   _____

3. Briefly list three basic things you know about hurricanes beyond what is explained on this page.

   _____

   _____

   _____

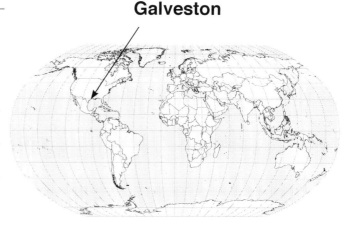

Galveston

Understanding Hurricanes
Activity Sheet 16B

Name _____ Class _____ Date _____

FOCUS *Read the information in the Fact Box and study the map. Then answer the questions.*

Fact Box: Hurricanes

Hurricanes are the most powerful storms in the world. They are severe cyclones (circular winds) that develop when winds in the warm, wet atmosphere of the tropics begin to circle around an area of low pressure. When the winds are less than 31 miles per hour, the storm is called a *tropical depression*. When they reach speeds of 32 to 74 miles per hour, the storm is called a *tropical storm*. When the winds exceed 74 miles per hour, the storm is classified as a *hurricane*.

- **Formation** As water evaporates from the ocean, it condenses into water droplets. This process releases heat energy, which warms the air and causes it to rise. Cooler surrounding air rushes in to replace it, causing winds. Given the right circumstance—very warm water, very warm and moist air, and a certain combination of winds—this process can cause the winds around a low-pressure area to spiral, creating a tropical depression and eventually a hurricane.

- **Structure** Hurricanes are circular patterns of clouds releasing huge amounts of rainfall and generating wind speeds of up to 200 miles per hour. (These winds rotate counterclockwise in the Northern Hemisphere and clockwise in the Southern Hemisphere). The low-pressure center of a hurricane is called the *eye*.

- **The Eye** The center of the hurricane is called the eye. Hurricanes swirl around this tiny (about 20 miles across) calm area in the center of the storm.

- **Storm Surge** The winds of a hurricane combine with the low pressure of the eye to create a "mound" of higher ocean water under the eye. In shallow water near land, this bulge of water is pushed up by the ocean floor, and a huge amount of water, called the *storm surge*, rushes inland, causing extensive damage.

- **Name** The word *hurricane* comes from the West Indian word *huran* meaning "big wind." The word *typhoon*, as hurricanes in the Pacific Ocean are called, comes from a Chinese word *taifun* meaning "great wind."

1. Where in the world do hurricanes originate? Based on the information in the Fact Box, explain why this is where they form.

_____

_____

_____

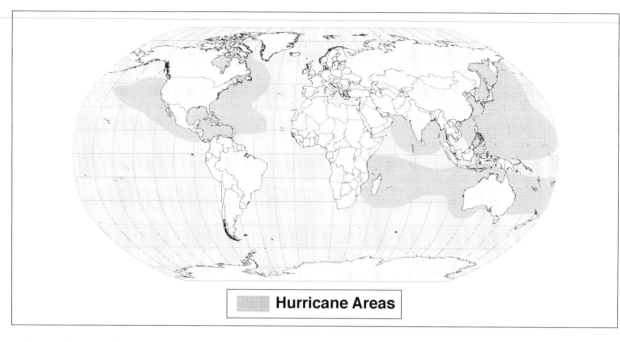

Hurricane Areas

2. Most damage from hurricanes comes from the storm surge rather than from powerful winds. What is a storm surge?

3. What is the origin of the word *hurricane*?

4. Why do you think hurricanes spiral in different directions in the northern and southern hemispheres?

5. How are air pressure and air temperature related to the formation of hurricanes?

Tracking a Hurricane
Activity Sheet 16C

Name _____ Class _____ Date _____

FOCUS *Every hurricane is given a name and is carefully tracked by meteorologists. Since the location, speed, and direction of each hurricane is regularly broadcast, you can track a hurricane along with the pros. Use the official hurricane symbol ( ) to identify the location of the hurricane on the map each day. Also, draw an arrow to show its direction of movement, and jot down the speed at which it is moving. Try to predict where the hurricane will go, along with if, where, and when it will hit the coast.*

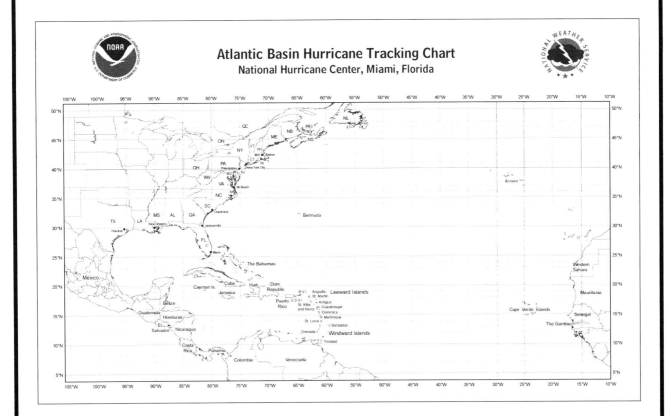

17. The World's Worst Tornado
The Great Tri-State Tornado

Objectives
- Identify and discuss the world's worst tornado
- Investigate how and why tornadoes form
- Demonstrate appropriate tornado safety procedures

Time Required
- About one class period

Teaching Tips
You may wish to introduce the topic by completing the first activity sheet as a class. The two additional activity sheets can be completed by students working independently, with partners, or in small groups.

Activity Sheet 17A
- Check student comprehension of the terms in the *Meteorologist's Handbook*; you may have students clip the entries from the *Meteorologist's Handbook* and the *World Weather Records Book* and add them to their compilations.
- Focus students' attention on the map.

Activity Sheet 17B
- Explain that tornadoes remain one of the "great unknowns" to meteorologists. We are still learning about their formation and their inner workings.
- Ask students whether and why they would like to become storm chasers.

Activity Sheet 17C
- You may wish to conduct a tornado drill during this class period.
- Contact the local National Weather Service officer or American Red Cross chapter to learn about specific tornado safety procedures.

Answers
- **Activity Sheet 17A** 1. about 62.5 miles per hour; 2. Answers will vary. Reward thoughtful and expressive responses.
- **Activity Sheet 17B** 1. Scientists who follow tornadoes to observe them at close range; 2. Check diagrams depicting tornadoes. Ensure they depict the interaction of warm and cool air.
- **Activity Sheet 17C** Examine the chart on tornado safety, and go over it with your class.

Extension and Enrichment
- You may wish explain the term *vortex* to students and demonstrate a vortex by swirling liquid in a bottle.
- The recent proliferation of video cameras has resulted in unprecedented, breathtaking footage of tornadoes. Show a collection of these images to the class.

Visit WorldRecordsBooks.com for more images and activities!

The World's Worst Tornado
Activity Sheet 17A

Name _____ Class _____ Date _____

FOCUS *Read the entries from the* Meteorologist's Handbook *and from the* World Weather Records Book. *Compare what you read with any personal experiences you have had with this type of weather. Then answer the questions that follow.*

From the *Meteorologist's Handbook*

- **tornado:** an extremely violent cyclone (rotation of winds) in the form of a narrow column of winds traveling up to 300 miles per hour and reaching from a thundercloud to the ground.

From the *World Weather Records Book*

The **world's worst tornado**, the Great Tri-State Tornado, traveled from Missouri, through Illinois, and Indiana on March 18, 1925. It stayed on the ground longer (three and a half hours) and traveled farther (219 miles) than any other tornado. It was also larger than any other, with a funnel width of one mile. This was also among the fastest-moving tornadoes ever seen. It killed 689 people.

1. The killer tornado of 1925 traveled 219 miles in three and a half hours. On average, how fast did it travel?

   _____

2. Describe any experience you've ever had with a tornado, either first-hand or through the media.

   _____

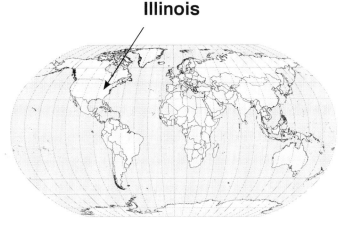

SunflowerEducation.net   Amazing World Records • Weather   **89**

Why Are There Tornadoes?

Activity Sheet 17B

Name _____ Class _____ Date _____

FOCUS *Tornadoes occur all over the world. When they form over oceans, they're called* **waterspouts.** *But most tornadoes occur in the United States. Why? Because the United States is a perfect "breeding ground" for tornadoes, providing all the elements needed for these storms to form. Read the article below to discover how tornadoes form. Then answer the questions.*

How Tornadoes Form

Meteorologists still don't know exactly how tornadoes form. Brave scientists called *storm chasers* follow tornadoes to observe them at close range and analyze them with scientific instruments.

Meteorologists know that most tornadoes form along fronts. A *front* is the place where air masses of different pressures, temperatures, and water content run into each other. At a front, the warm air rises above the cool air. Cool air rushes in to replace it creating winds. Sometimes, fast-moving winds travel over slower winds causing the air to spin and form a horizontal tube of rapidly rotating air at the base of a huge cloud. A certain combination of winds can cause one end of this tube to dip toward the ground, creating a tornado.

1. Who are *storm chasers*?

   _____

   _____

2. Sketch a diagram to show your understanding of how a tornado forms.

Tornado Safety
Activity Sheet 17C

Name _____ Class _____ Date _____

FOCUS *More tornadoes occur in the United States than in any other country on Earth. So you need to know what actions to take in the event of a tornado. The chart contains important—and potentially lifesaving—information. Study it carefully, answer the questions, and memorize what you learn.*

Tornado Safety	
If You're Under a Tornado Watch A tornado watch means that conditions are right for the creation of tornadoes • Remain alert. Keep an eye on the weather. Tune in to television or radio weather broadcasts frequently for the latest information. • Rehearse your Plan of Action	How I will know if there is a tornado watch: Where I would obtain information during a tornado watch:
If You're Under a Tornado Warning A tornado warning means that a tornado has actually been spotted by persons on the ground or by radar. • Immediately implement your plan of action. • Seek shelter. Keep an eye on the weather. Tune in to television or radio weather broadcasts frequently for the latest information.	How I will know if there is a tornado warning: **Plan of action:** Where I will go if there is a tornado warning: • at home: • at school:

18. The World's Worst Flood
The Huang Ho River Flood of 1882

Objectives
- Identify and discuss the world's worst flood
- Explain why the Huang Ho River is so prone to floods
- Explore human attempts to protect themselves against floods

Time Required
- About one class period

Teaching Tips
You may wish to introduce the topic by completing the first activity sheet as a class. The two additional activity sheets can be completed by students working independently, with partners, or in small groups.

Activity Sheet 18A
- Check student comprehension of the terms in the *Meteorologist's Handbook*; you may have students clip the entries from the *Meteorologist's Handbook* and the *World Weather Records Book* and add them to their compilations.
- Focus students' attention on the map.

Activity Sheet 18B
- Make sure students understand that heavy rainfall in the regions drained by a river's tributaries is the major reason for river floods.
- Explain that there are other types of flooding, such as urban flooding and coastal flooding.

Activity Sheet 18C
- Point out that these measures fall short of completely protecting people from floods.
- Quiz students about what they know about personal flood safety.

Answers
- *Activity Sheet 18A* 1. Check the maps to ensure the 1887 flood is correctly drawn; 2. Answers will vary, but should include that many floods are caused by heavy precipitation. Reward thoughtful responses.
- *Activity Sheet 18B* 1. because it has flooded so often, for so long, and with such devastation; 2. The sediments settling to the bottom make the river shallower and easier to flood; 3. great amounts of rainfall upstream and on a river's tributaries; 4. because the area that is flooded is very fertile.
- *Activity Sheet 18C* Encourage students to visit the web to conduct research on different methods of defense against floods. Reward thoughtful responses.

Extension and Enrichment
- Have students research and report on the Midwest Flood of 1996.
- Explain what a flash flood is and instruct students on flash flood safety.

Visit WorldRecordsBooks.com for more images and activities!

The World's Worst Flood
Activity Sheet 18A

Name _____ Class _____ Date _____

FOCUS *Read the entries from the* Meteorologist's Handbook *and from the* World Weather Records Book. *As you read, compare what you are reading with any personal experiences you have had with this type of weather. Then complete the activities that follow.*

From the *Meteorologist's Handbook*
- **flood:** an overflowing of water that covers land that is normally dry
  The term *flood* usually refers to river floods.

From the *World Weather Records Book*
The **world's worst flood** occurred in September and October, 1887, along the Huang Ho River in China. The river overflowed its banks, covering approximately 50,000 square miles with water, destroying 300 villages, and killing 2.5 million people.

1. How do you think floods are related to the weather?

   _____
   _____
   _____
   _____

2. What do you think people have done to protect themselves from floods?

   _____
   _____
   _____

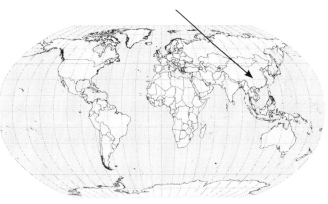

The River of Sorrow
Activity Sheet 18B

Name _____ Class _____ Date _____

FOCUS *The worst flood in history occurred along the Huang Ho River in China. To understand why this disaster occurred, read the article. Then answer the questions that follow.*

Nearly 3,000 miles long, the Huang Ho (also spelled Huang He) is the second-longest river in China, and one of the longest rivers in the world. You may know it as the Yellow River. This name comes from its yellowish-brown color, a result of the millions of tons of yellow sediment that the river is constantly carrying downstream.

Its great length and unusual color are enough to make this river famous the world over. But the Huang Ho is not so much famous as it is infamous, or notorious. This river has flooded so often, for so long, and with such devastating results that it is known by the name "China's Sorrow" or "The River of Sorrow."

The flooding is largely a result of the sediments that give the river its name. As the sediments settle, they make the bed of the river's channel shallower. Because the river is so shallow, it floods more easily. When the river swells from great amounts of rainfall upstream and on its tributaries—the conditions that cause most floods worldwide—the river channel can't handle all the extra water. The water then spills over the plain through which the river runs, with terrible results.

This process has repeated itself countless times. For centuries, the Chinese have tried to stop the flooding by building levees along the river. But the waters often prove too much for the levees, breaking through or spilling over them, and the sorrow continues.

When the Huang Ho does flood, it's particularly devastating because the plane through which the river runs is highly populated, mostly by poor farmers. Why? Because the very sediments that contribute to the flooding also make excellent farmland. Nobody knows how many millions of people have died in the floodwaters of the River of Sorrow.

1. Why is the Huang Ho called "China's Sorrow"?

   _____

   _____

2. How do sediments contribute to flooding on the Huang Ho?

   _____

   _____

3. Besides sedimentation, what two factors combine to create floods on the Huang Ho, and on rivers worldwide?

_____

_____

_____

4. Why do so many people live near the Huang Ho?

_____

_____

_____

_____

The Huang Ho River.

Controlling Floods
Activity Sheet 18C

Name _____ Class _____ Date _____

FOCUS Because floods can be so damaging—and so deadly—people have tried to stop them, or at least control them, for thousands of years. These efforts continue today. To learn about them, conduct research and complete the chart.

How People Defend Against Floods

Human Construction	What It Is	How it Protects People Against Floods
Dam		
Levee		
Flood Wall		
Dike		
Sea Wall		

19. The World's Worst Weather Disaster
Cyclone of 1970

Objectives
- Identify and discuss the world's worst weather disaster
- Explain the factors that combine to make the Bay of Bengal the site of the world's deadliest weather
- Identify and prepare for severe weather in their community

Time Required
- About one class period

Teaching Tips
You may wish to introduce the topic by completing the first activity sheet as a class. The two additional activity sheets can be completed by students working independently, with partners, or in small groups.

Activity Sheet 19A
- Check student comprehension of the term in the *Meteorologist's Handbook*; you may have students clip the entries from the *Meteorologist's Handbook* and the *World Weather Records Book* and add them to their compilations.
- Focus students' attention on the map.

Activity Sheet 19B
- Guide students in understanding how geographic, weather, and economic factors combine to create the weather tragedies in the Ganges-Brahmaputra Delta.
- Foster student appreciation of the weather information available in the United States.

Activity Sheet 19C
- Take care to avoid frightening students.
- Conduct appropriate drills in conjunction with this activity.

Answers
- **Activity Sheet 19A** 1. about one million people; 2. a huge cyclone.
- **Activity Sheet 19B** 1. a. weather factors: very powerful cyclones strike this area, b. geographic factors: the funnel shape of the bay focuses the high waters, the very low land is quickly flooded; c. economic factors: the people of the region are too poor to afford radios or televisions that would give them warning of the approaching storm; 2. Answers will vary. Reward thoughtful responses.
- **Activity Sheet 19C** Guide students as they complete the activity sheets.

Extension and Enrichment
- Have students locate contemporary newspaper reports of the tragedy and share them with the class.
- Have students report on other, more recent cyclones in the region.

Visit WorldRecordsBooks.com for more images and activities!

The World's Worst Weather Disaster
Activity Sheet 19A

Name _____ Class _____ Date _____

FOCUS *Read the entries from the* Meteorologist's Handbook *and from the* World Weather Records Book. *As you read, compare what you are reading with any personal experiences you have had with this type of weather. Then answer the questions that follow.*

From the *Meteorologist's Handbook*
- **natural disaster:** a catastrophe bringing great loss of human life and great damage that is caused by a natural event, such as severe weather, a flood, an earthquake, a volcanic eruption, etc.

From the *World Weather Records Book*
The **world's worst weather disaster,** in terms of human lives lost, for which solid figures are available, occurred in Bangladesh on November 12 and 13, 1970. A huge cyclone, or hurricane-like circular storm, slammed into the coast, flooding low-lying areas and killing about one million people.

1. How many people died in the world's worst weather disaster?
   _____

2. What type of weather caused this disaster?
   _____

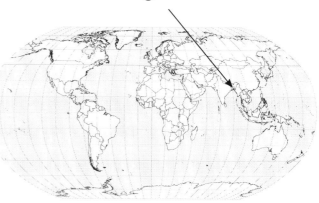

98 *Amazing World Records • Weather* SunflowerEducation.net

Natural Disasters at the Head of the Bay of Bengal
Activity Sheet 19B

Name _____ Class _____ Date _____

FOCUS — *Some people think that the Bay of Bengal earns a world record as the location of the "deadliest weather on Earth." They may be right. To develop your own opinion on the matter, read the article, and answer the questions.*

> The Ganges-Brahmaputra Delta is the most densely populated area on Earth. About 300 million people—a population greater than that of the United States—live in the delta region, an area about as large as Texas.
>
> Tragically, most of these people are extremely poor. Tragic, too, is the fact that they live in an area that is home to some of the most devastating weather on Earth.
>
> The Ganges-Brahmaputra Delta is located on the Bay of Bengal in the Indian Ocean. Cyclones, as hurricanes in this part of the world are called, are born over the Indian Ocean and travel northward into the bay. The high waters of the storms squeeze into the funnel shape of the bay, making them higher still. By the time they reach the head of the bay, where millions of Indians and Bangladeshis live, the waters sweep over the land in waves as high as 50 feet.
>
> The land here is very low—almost at sea level. The poor farmers largely live on *chars*, low islands created by sediment deposited in the delta. When cyclones, with their high waters, rush up the Bay of Bengal, the chars are quickly and completely flooded. A terrible loss of life results. The people of the region are too poor to afford radios or televisions that would give them warning of approaching storms and, in many cases, they lack the means to escape anyway.

1. Identify how factors combine to make the head of the Bay of Bengal the site of the world's worst weather disasters.

   a. weather factors:

   _____
   _____
   _____
   _____
   _____
   _____

b. geographic factors:

_____
_____
_____
_____
_____
_____

c. economic factors:

_____
_____
_____
_____
_____
_____

2. What do you think could be done to help the people of this region?

_____
_____
_____
_____
_____
_____
_____
_____

Preparing for a Natural Disaster
Activity Sheet 19C

Name _____ Class _____ Date _____

FOCUS — *No matter where you live in the world, there's a chance that a weather-related natural disaster will strike your area. This is especially true in the United States, where severe weather is common. The keys to dealing with severe weather are knowledge and preparation. Get both keys now by completing the boxes, with guidance from your teacher.*

Severe Weather That Might Strike My Community

How I Can Prepare for Severe Weather and Protect Myself Against It

20. The World's Worst Weather
The United States

Objectives
- Identify and discuss the country with the world's worst weather
- Explain why there is so much severe weather in the United States
- Summarize the purpose and activities of the National Weather Service

Time Required
- About one class period

Teaching Tips
You may wish to introduce the topic by completing the first activity sheet as a class. The two additional activity sheets can be completed by students working independently, with partners, or in small groups.

Activity Sheet 20A
- Check student comprehension of the terms in the *Meteorologist's Handbook*; you may have students clip the entries from the *Meteorologist's Handbook* and the *World Weather Records Book* and add them to their compilations.
- Focus students' attention on the map.

Activity Sheet 20B
- Take care to avoid frightening students.
- Emphasize that preparation and information are keys to staying safe during severe weather.

Activity Sheet 20C
- Make sure that students understand that the National Weather Service is a federal government agency.
- Emphasize how vital weather knowledge and forecasts are to all areas of life.

Answers
- *Activity Sheet 20A* 1. Answers will vary. Reward thoughtful responses; 2. Answers will vary. Provide the students with a safe environment to discuss potentially difficult memories. Reward thoughtful responses.
- *Activity Sheet 20B* 1. The United States stretches from tropical areas to the Arctic. It is bordered by oceans, where hurricanes form. It is a meeting ground for dry, cold air from the north and warm, moist air from the south resulting in thunderstorms; 2. Answers will vary. Reward thoughtful responses.
- *Activity Sheet 20C* 1. because the weather affects everyone; 2. Answers will vary and may include the following responses: closing schools, repairing power lines and when to plant crops. Reward thoughtful responses; 3. Answers will vary and may include the following responses: meteorologist, climatologist, storm chaser, statistician, and forecasters. Reward thoughtful responses.

Extension and Enrichment
- Explain what a weather radio is and keep one in the classroom.
- Invite a representative from the National Weather Service to speak to students.

Visit WorldRecordsBooks.com for more images and activities!

The Country with the World's Worst Weather
Activity Sheet 20A

Name _____ Class _____ Date _____

FOCUS *Read the entries from the* Meteorologist's Handbook *and from the* World Weather Records Book. *As you read, compare what you are reading with any personal experiences you have had with this type of weather. Then answer the questions that follow.*

From the *Meteorologist's Handbook*
- **meteorology:** science concerned with the study of weather and climate

From the *World Weather Records Book*
The country with the **world's worst weather** is the United States of America.

1. Does it surprise you that the United States has the world's worst weather? Why or why not?

   _____

   _____

   _____

   _____

2. What firsthand experiences do you have to indicate that you live in a country with the world's worst weather?

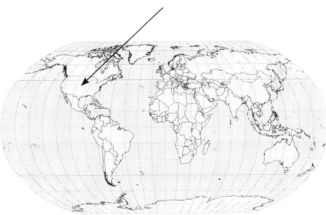

   _____

   _____

   _____

   _____

   _____

   _____

SunflowerEducation.net — Amazing World Records • Weather — 103

Understanding Weather in the United States
Activity Sheet 20B

Name _____ Class _____ Date _____

FOCUS *The United States has the world's worst weather. But why? To find out, read the article. Then answer the questions.*

Severe Weather in the United States

The United States has more extreme weather than any other country on Earth. In any given year, Americans in different parts of the country will experience tornadoes, floods, droughts, hurricanes, blizzards, severe thunderstorms, temperatures well above 100 degrees, temperatures will below 0 degrees, and more.

Why is the weather here so wild? One of the reasons is that the United States is a large country. It stretches from tropical areas to the Arctic. Another reason is that it is bordered by oceans, where hurricanes form. A third, very important reason is that the United States is a meeting ground for cold, dry air from the north and warm, moist air from the south. When these air masses collide, violent thunderstorms and tornadoes are created across much of the middle of the country.

1. List three reasons why weather in the United States is so extreme.

   _____
   _____
   _____
   _____
   _____
   _____

2. What kinds of severe weather do you experience in your part of the country?

   _____
   _____
   _____
   _____
   _____

104 *Amazing World Records • Weather*   *SunflowerEducation.net*

The National Weather Service
Activity Sheet 20C

Name _____ Class _____ Date _____

FOCUS *Most of our knowledge of weather in the United States comes from the National Weather Service, an agency of the federal government. Read the following excerpt, which was written by officials at the National Weather Service. Then answer the questions.*

Who We Are

The National Weather Service is the primary source of weather forecast and warning information in the United States. Television weather forecasters and private meteorology companies prepare their forecasts using basic forecast guidance issued daily and weather observations issued hourly by the National Weather Service.

Every day, millions of weather-based economic decisions are made in agriculture, transportation, power, construction, and other sectors of the economy. Information issued around the clock by the National Weather Service affects the life of every American.

1. Explain why the last sentence is true.

2. Give at least three specific examples of "weather-related decisions."

3. What types of jobs do you think members of the National Weather Service do?

Made in the USA
San Bernardino, CA
13 April 2017